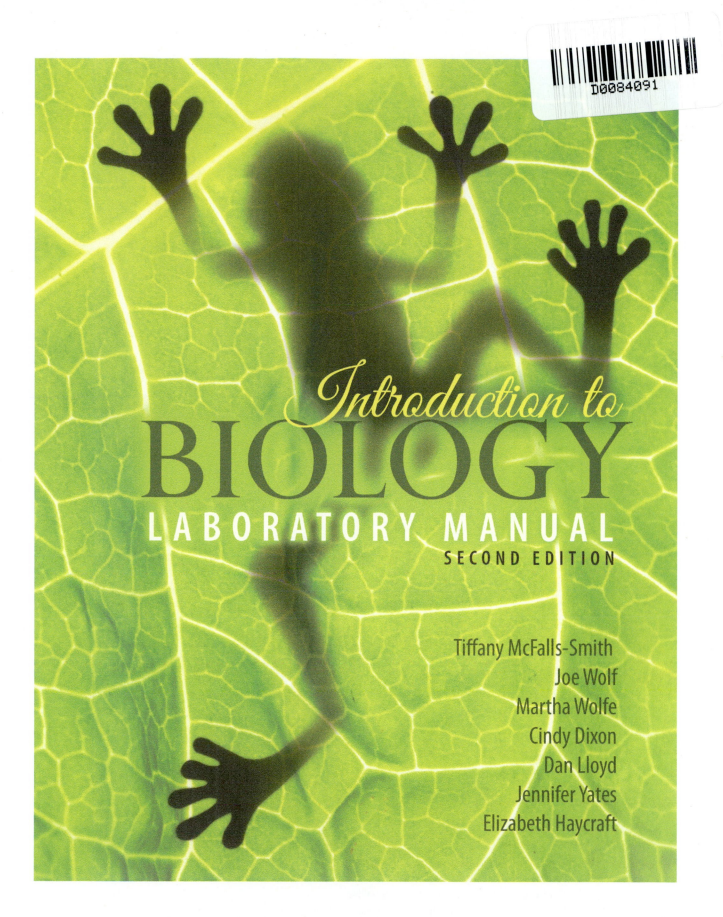

Introduction to
BIOLOGY
LABORATORY MANUAL
SECOND EDITION

Tiffany McFalls-Smith
Joe Wolf
Martha Wolfe
Cindy Dixon
Dan Lloyd
Jennifer Yates
Elizabeth Haycraft

Kendall Hunt
publishing company

www.kendallhunt.com
Send all inquiries to:
4050 Westmark Drive
Dubuque, IA 52004-1840

Published in the United States of America

Laboratory Activities

Laboratory Rules

This laboratory is a place to perform assigned experiments during scheduled class time. Read the laboratory procedure before you arrive in the lab. Preparing for the lab allows you to formulate a plan, work more efficiently, and gain more understanding from the laboratory activities.

The laboratory may be available for study time; please check with your instructor or the laboratory coordinator (270–706–8543) if the laboratory is needed outside of scheduled class time. Models, slides, and dissection materials may not be removed from the lab.

THE LABORATORY CAN BE A DANGEROUS PLACE WHEN:

1. *YOU ARE CARELESS.* Read your procedure and formulate a plan.
2. You misuse equipment.
3. You use improper laboratory technique.
4. You fail to ask for assistance when you do not understand.
5. Pathways are cluttered with books, book bags, and personal property.

IT IS IMPERATIVE:

1. That you respect the safety of others.
2. That you follow directions given by the instructor.
3. That you understand all warnings and special instructions.
4. That you follow special procedures.

SAFETY EQUIPMENT:

1. Safety goggles must be worn, <u>over your eyes</u>, during all chemistry labs.
2. Emergency showers are located in 203A, 202, and 202B.
3. Emergency eye wash stations are located in 203A, 202, and 202B.
4. The fire extinguisher is located at the front of the room.
5. The emergency gas shut-off valve is located at the front of the room.
6. Closed toed shoes must be worn <u>during all labs.</u>

BROKEN EQUIPMENT:

1. All injury or equipment breakage must be reported to the instructor, no matter how insignificant it may seem at the time of occurrence.
2. Even the most experienced lab professional can break glass equipment. Take extra care when using glassware, particularly if wearing gloves.
3. Use proper clean-up and disposal procedures. Ask for guidance when needed.
4. All broken glass will be disposed of in the special container marked "Glass". **DO NOT** put paper and other waste in the "Glass" container. **DO NOT** put broken glass in the garbage cans.
5. Broken equipment will be charged to the student.

FOOD AND DRINKS:

1. Food and drinks are NOT allowed in the lab.

MAINTENANCE:

1. Chemistry boxes are shared with multiple classes; therefore, it is your responsibility to leave the glassware in the boxes clean. Chemical residual from your experiment may interfere with the next classes experiment. If you fail to clean your chemistry equipment, points may be deducted from your lab grade.
2. All glassware and utensils need to be washed with soap and water after usage. The items must be dried thoroughly and returned to the chemistry box. Beakers and test tubes must be inverted during storage.
3. Dissecting utensils must be washed with soap and water and dried and placed in tray and returned to the drawer.
4. Dispose of all solid waste in the appropriate container.
5. Dispose of Hazardous Materials as directed.

LAB ACCESS:

1. Participants in campus sanctioned events and students enrolled in relevant ECTC courses are permitted in laboratory spaces under the supervision of authorized personnel only.
2. Other campus visitors, particularly children, are not allowed in the labs due to OSHA regulations.

Chemistry Equipment Inventory

Lab Box Number: _____

Name: _____

Lab partners: _____

_____

Number of Items/Begin	Cost per Item	Number of Items/End	Number of Items/Begin	Item/Cost	Number of Items/End
	(5) 250ml Beakers $4.25			Test Tube Clamp $10.00	
	400ml Beaker $5.00			Large Rack $10.00	
	Evaporating Dish $6.00			Test Tube Brush, Small $3.00	
	10ml Graduated Cylinder $6.00			Test Tube Brush, Large $3.00	
	100ml Graduated Cylinder $10.00			Ruler $1.00	
	(10) Large Test Tubes $1.00			Funnel $3.00	
	(10) Small Test Tubes $1.00			Forceps $5.00	
	(2) Stirring Rods $3.00			Scoopula	
				$5.00 (2)250mlErlenmeyer Flask $5.25	

Agreement: I have been given a copy of the safety rules and guidelines. I have read the rules and guidelines. I understand the rules and guidelines. I agree to comply with all the rules and guidelines while enrolled in this lab. **I understand that points will be deducted if I fail to comply with safety rules and guidelines**. Please sign and date.

Signature:_____/Date_____

LAB 1

The Nature of Science

Science is just one way of knowing the world. Some people experience the world through the lens of music, others are passionate about history. It's been said that those who grow up to be scientists have simply never outgrown the childhood phase of awe and experimentation that we can all remember.

WHAT IS SCIENCE AND HOW IS IT DONE?

The objective of this lab is to get you to think about the nature of science, the connection between science and society, and how you can adopt the lens of science throughout this course and, hopefully, throughout your life. Part of the fun of science is the discovery of new things. Harvard's Edward O. Wilson, a myrmecologist (specialist in ants), is known for several things, including two Pulitzer prizes. During an interview at the age of 78, the interviewer noted, "You've spent your entire career doing what you did when you were nine years old." Wilson's reply: "I get the same thrill. . . . There is no better high than discovery."

Hopefully, in lab you will be led on a path of discovery. Of course, labs are designed for you to discover what we already think we know and understand about nature. You are repeating experiments that have been done many times before, but you need to grasp the basic principles of science so that you can recognize real science from ideas (no matter how passionately expressed) that are nothing more than science fiction. So when do we know it's science?

The word "science" comes from the Latin word *scientia* meaning "knowledge," and that is what we've come to depend on. This endeavor called science that gives us a body of organized, testable, reliable knowledge also gives us a logical and rational explanation about the world around us. Every time we take an aspirin, turn on a light switch, or start our car, we rely on that knowledge and our current understanding of how the natural world works. Do we know all there is to know? Of course not. Is what we think we know entirely correct? Probably not. But, the more testing we do and the more evidence we gather bring us closer to an understanding of the mechanisms working in the world around us.

Science is an evidence-based endeavor. Consequently, what you believe or want to be true doesn't make it so. Of course, scientists are just as human as anyone, and so have biases. They try to recognize and account for their biases. If not, someone else will do it for them!

That's why science is done in a community context. This means that the tests and evidence are published for all to see so that any biases can be discovered and eliminated.

On the other hand, science is *not* a democracy. Just because many people might want something to be true and would like it to be taught as an alternative to science does not make it science. Wouldn't it be nice if we could pick up the newspaper and every time our horoscope said it would be our lucky day, go out and buy a lottery ticket with the surety that it was a winner! But when we are sick and go to the doctor, we count on something more concrete than luck. We depend on the knowledge, gained by testing and evidence gathering, to be as close to correct as possible.

Adapted from *Biology 101 Lab Manual* by Ann S. Evans, Sarah Finch, Eric Lamberton and Randy L. Durren. Copyright © 2014 by Kendall Hunt Publishing Company. Reprinted by permission.

There is an order to science and the process of science. Without this order—this process—there is only conjecture, speculation, and guesses. It's okay to have new ideas and speculate how things work, but science needs to continually test every assumption. Those things that can't be tested must remain outside the realm of science. So, what is this process?

You learned *The Scientific Method* in school:

1. Observe: Notice something!
2. Question: Wonder why or how or when . . .
3. Hypothesize: Speculate; come up with a reason.
4. Predict: Come up with a "What if?" or "If . . . then . . ."
5. Test: Collect data.
6. Analyze: Make sense of the numbers.
7. Decide: Do your data support your hypothesis or not?
8. Communicate: Put your contribution in context.

Do scientists think *The Scientific Method* describes what they do? Yes . . . but doing science is a bit more complicated than that linear prescription. How science really works looks more like this.

There's No Such Thing as Scientific Proof

Testing ideas is a huge part of doing science. However ideas are not tested directly. What scientists test are the predictions that follow from a particular idea. If the data fit the predictions, the hypothesis is supported. Supporting a hypothesis is not the same as proving a hypothesis. In fact, there is no such thing as scientific proof (except in the sense of a mathematical proof). Even from a philosophical perspective, you can't prove anything positively.

Ads for various products often claim: Scientifically proven to. . . . What is wrong with this claim?

_____

_____

Ideas Have to Be Not Only Testable, but Falsifiable

If evidence does not support a prediction—if it contradicts a prediction and therefore the hypothesis from which the prediction arose—then we say the hypothesis has been rejected. It has been proved false, or falsified.

Predictions that follow from hypotheses must be falsifiable: able to be falsified. If it is not possible to conceive of a case in which the prediction could be proved false, then the idea cannot be tested, and therefore cannot be addressed by science.

Give an example of an idea or question that cannot be addressed by science.

_____

_____

Religion Is Not within the Purview of Science, but That Does Not Mean That Science and Faith Are Mutually Exclusive

Religion is by definition a matter of faith and matters of faith are not subject to test. Therefore, religious questions are not amenable to scientific inquiry. The two fields cannot make contributions to one another. But that does not mean that they are mutually exclusive. In western countries, especially the United States, scientists are likely to describe themselves as religious or spiritual.

What is wrong with this statement: You can't believe in both God and evolution.

_____

_____

It's Not "Just a Theory" It's a Scientific Theory

All fields have jargon or special meanings of common words. In the vernacular (lay, or common usage), the word *theory* means an opinion: "My theory is that they divorced (or she was promoted or he said no) because. . . ." But in science, the word *theory* means something very different. A theory is an idea or conceptual framework that has been tested repeatedly and has yet to be proven false. In other words, the preponderance of the evidence jibes with the theory, and therefore supports it. Examples include the theory of gravity, the germ theory of infectious diseases, the theory of plate tectonics, the theory of evolution. To say "It's just a theory" about a scientific theory confuses the lay and scientific uses of the word.

Explain the difference between the lay and scientific uses of the term theory.

_____

_____

Theory versus Fact

Theories cannot be positively proved. Can they be disproved? Of course, new evidence can come to light; the "facts" may change. Definitions of the terms *fact* and *theory* were perhaps best put by Stephen J. Gould in the context of the theory of evolution:

Well evolution *is* a theory. It is also a fact. And facts and theories are different things, not rungs in a hierarchy of increasing certainty. Facts are the world's data. Theories are structures of ideas that explain and interpret facts. Facts don't go away when scientists debate rival theories to explain them. Einstein's theory of gravitation replaced Newton's in this century, but apples didn't suspend themselves in midair, pending the outcome. And humans evolved from ape-like ancestors whether they did so by Darwin's proposed mechanism or by some other yet to be discovered.

Moreover, "fact" doesn't mean "absolute certainty"; there ain't no such animal in an exciting and complex world. The final proofs of logic and mathematics flow deductively from stated premises and achieve certainty only because they are *not* about the empirical world. Evolutionists make no claim for perpetual truth, though creationists often do (and then attack us falsely for a style of argument that they themselves favor). In science "fact" can only mean "confirmed to such a degree that it would be perverse to withhold provisional consent." I suppose that apples might start to rise tomorrow, but the possibility does not merit equal time in physics classrooms.

Stephen J. Gould, ("Evolution as Fact and Theory"; *Discover*, May 1981)

Can facts be disproven? Why or why not?

_____

_____

EXPERIMENTAL DESIGN

Even though experimentation is not the only way to test ideas, it is an important way. Solidify your understanding of relevant terms by working through the following examples (exerpted from www.LessonPlansInc.com). For more information see your textbook by Hoefnagels.

> **Independent variable:** The one that the investigator manipulates in order to ask whether the value of this variable is related to change in another variable (dependent variable).

> **Dependent variable:** The one that is measured to answer the question: Does the value of this variable change when the value of the independent variable changes? (Does it depend on the value of the independent variable?)

> **Standardized variables:** Ones that the investigator purposefully holds constant for all test subjects, so that as much as possible, the independent variable is the only one that influences the outcome. Sometimes called **controlled variables**.

> **Control group:** The one that receives the "normal" or untreated value of the independent variable. Not all experiments have a clearly identified control group.

Identify the independent variable (IV), the dependent variable (DV), the control group (CG) and standardized variables (SV). Think about how the data might be plotted. Make a sketch to identify to the independent variable (on the x-axis) and the dependent variable (on the y-axis).

The number of flowers on different breeds of rose bushes in a greenhouse is recorded every week for two months.

IV _____

DV _____

CG _____

SV _____

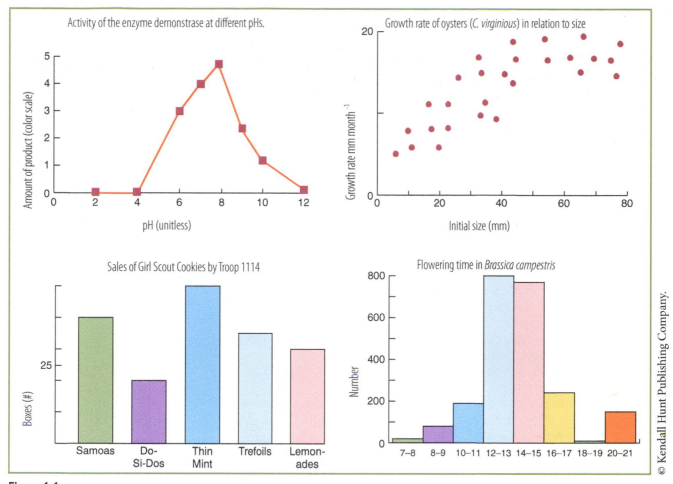

Figure 1.1

You notice that the shower is covered in a strange pink slime. You decide to try to get rid of this slime by adding lemon juice. You spray half the shower with lemon juice and spray the other half of the shower with water. After 3 days of treatment, there is no change in the appearance of the pink slime on either side of the shower.

IV _____

DV _____

CG _____

SV _____

One tank of goldfish is fed once a day, a second tank twice a day, and a third tank four times a day during a six-week study. The fishes' body fat is recorded.

IV _____

DV _____

CG _____

SV _____

Graphs
How to Make a Graph
It is much easier to comprehend quantitative data in graphical form. But poorly constructed graphs can be difficult to understand or even misleading. Every graph should have the following features:

- Title
  - Should describe what the graph shows
    - Example: *The relationship between texting and car crashes* rather than *Texting and driving*
- Appropriate placement of variables
  - In a manipulative experiment
    - Independent variable on the x-axis
    - Dependent variable should be on the y-axis
- Axis labels
  - The name of the variable
    - Example: *Breath-holding ability* rather than *Breathing*
- Axis units
  - Be precise
    - Examples: for time, seconds, minutes, etc.; for temperature, °C or °F
- Appropriate scale for axes
  - Base the scale on the minimum and maximum values of the variable
    - Example: If range of values is 0–5, scale should not go to 10
  - Starting each axis with 0 may not be appropriate
    - Example: Range of values 2000–2600, don't start scale at 0
  - The scale must be uniform
    - One unit must be of the same length along the axis; use a ruler
- Legend
  - If more than one symbol is used
    - Example: *The effect of temperature on germination time in five species*
    - Indicate which symbols correspond to which species

Types of Graphs
Which type of graph to use depends upon your data. For some data sets, more than one type of graph would be appropriate.

- Line graphs
  - Data are plotted onto a grid as points and then connected
  - Often used for manipulative experiments with independent and dependent variables
- Scatter plots
  - Data are plotted onto a grid as points, but not connected
  - Useful for visualizing patterns in descriptive data
- Bar graphs
  - Used for categorical data
  - Each bar shows the value for a different category
  - Since categories are discrete, bars are separated by space
- Frequency histograms
  - The height of the bar corresponds to the number of individuals that show a particular value (or range of values) for the variable
  - Since the variable (e.g., height) is continuous, bars are touching

INTERPRETING DATA AND GRAPHING

Understanding data and making sense of information are key to the scientific process. Humans are visual creatures and because of that, we understand visual things and usually interpret the world around us visually. Graphing is a way of putting data in a visual format. It helps us "see" the relationships among pieces of data.

Here are the data from the first experiment example. For simplification, we have only included three of the varieties involved in the study.

Flower numbers listed from week 1 to week 15

Rose Variety A: 0, 0, 2, 12, 15, 16, 15, 12, 15, 20, 22, 25, 25, 30, 35

Rose Variety B: 0, 1, 6, 10, 22, 31, 32, 43, 44, 16, 6, 2, 2, 1, 1

Rose Variety C: 0, 2, 16, 36, 25, 22, 15, 13, 12, 24, 33, 36, 22, 16, 10

From looking at the data you might be able to make some basic conclusions, but if you were analyzing hundreds of varieties it becomes very difficult.

Graph the data shown on a simple line graph. Be sure to include all components: title, appropriate placement of variable, axis labels and units, appropriate scale and legend. Hint: Using different colors for each rose variety might give you what you want.

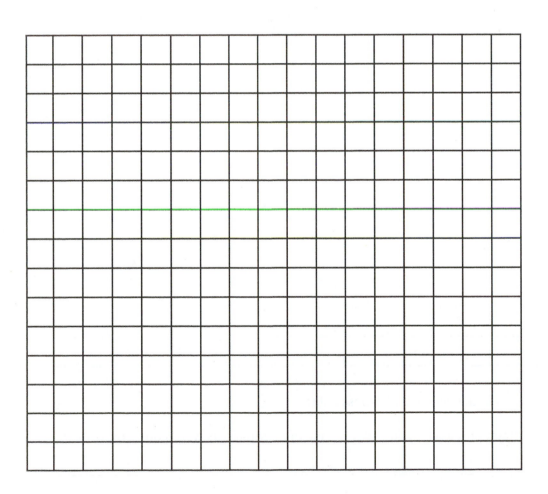

Now answer some basic questions using the graph.

Which variety has the most blooms in mid summer? _____

If you wanted to have blooms late into the fall, what variety might you pick? _____

Which variety seems to have two heavy bloom pick times? _____

Here are the data for the third experiment. Graph these data using a bar graph this time. There are six fish in each tank. Hint: How can you show the gist of the information without having to plot each number?

	Goldfish body fat (grams) at start	Goldfish body fat (grams) after 6 weeks
Tank #1	15.6, 11.2, 10.3, 14.6, 13.1, 12.4	15.8, 11.7, 11.0, 15.2, 13.5, 12.5
Tank #2	12.2, 13.7, 15.4, 16.7, 13.2, 14.2	12.5, 14.3, 15.9, 17.7, 14.1, 14.8
Tank #3	11.4, 11.6, 14.7, 12.6, 15.3, 13.7	11.6, 12.0, 15.6, 13.3, 15.9, 14.4

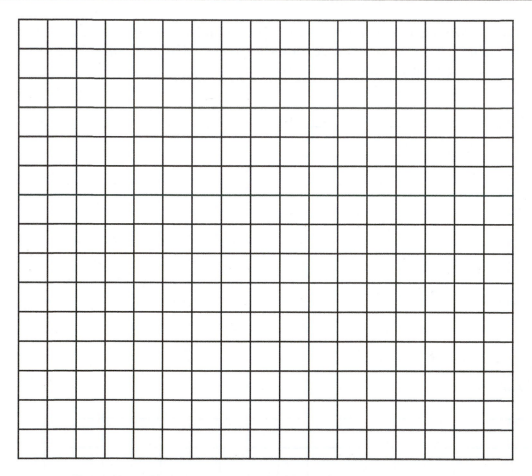

Can you draw any conclusions about these data? _____

Write a short statement about your next course of action if this were your experiment. _____

_____

_____

The Metric System

In the last exercise, the unit of measurement was the gram. This is the metric system, which is the accepted system throughout the world and therefore always used in science. Every country in the world has adopted the metric system as the worldwide standard for international trade, so even business majors should become familiar with this system. There are only a few countries in the world that have not completely converted to the metric system within the general population. Unfortunately, the United States is one of them. The metric system is not difficult; actually, it is easier than the English system, it's just unfamiliar to many people. The following exercise is designed to familiarize you with the basic units of measurement in the metric system.

We start with length. The basic unit of length for the metric system is the _____.

This unit is a little bit larger than a yard. The abbreviation is (m)_____.

The term used for 1,000 of something is the kilo, so 1,000 meters is called a _____.

This unit is a little bit larger than a ½ mile. The abbreviation is (km)_____.

We use the term for 1/100 of something all the time. For example, 1/100 of a dollar is called one

_____. (Take the copper coin out of your pocket and read what is says.) The abbreviation is (cm).

You should have a meter stick on your table. How many *numbered* marks are on a meter? _____.

Therefore each one of these marks is a _____.

Each cm is divided into _____ spaces (tiniest marks). 100 cm times 10 = 1,000.

The term used for 1/1,000 of something is milli. So each little mark is 1/1,000 of a meter or a

_____. The abbreviation is (mm).

Here is a cube 1 centimeter by 1 centimeter by 1 centimeter.

This cube, being 1 cm on each side is called a cubic _____ and the abbreviation is

(_____) or (_____).

This cube can be filled up. The space inside it is called volume.

It turns out that 1,000 of these volumes, these cubic centimeters, equals one liter, which is the basic metric measurement for volume. A liter is a little bit larger than a quart. The abbreviation is (l).

Remember from earlier, the term used for 1/1,000 of something is _____.

Therefore 1/1,000 of a liter is called a _____. The abbreviation would be

(_____).

It also turns out that 1 milliliter (ml) of water weighs 1 gram, which is the basic metric measurement of mass.

So, 1 cubic _____ (a measurement in length) = 1 _____ and

(a measurement in volume) = 1 _____ (a measurement of mass).

This simple conversion from length to volume to mass, without the use of a calculator or computer, cannot be done in the English system.

One last conversion, what is 1,000 grams called? _____ The abbreviation would

be (_____).

This _____ weighs a little bit more than 2 pounds.

SERENDIPITY AND SILLY PUTTY

A great example of the role of serendipity in science (and making money!) is Silly Putty. During World War II, a General Electric engineer named James Wright was trying to create a substitute for rubber. (The Japanese had cut off access to rubber, limiting production of tires, boots, etc.) Wright combined boric acid and silicone oil; the result was no good as a rubber substitute. Years later, after no practical use could be found for the material, it was marketed as an adult novelty toy.

Silly Putty is unusual because it acts as both an elastic solid and a viscous fluid. What gives it its unusual properties is the fact that it is an elastomer, which is a special kind of polymer. A polymer is a molecule made of many similar, repeating units called monomers.

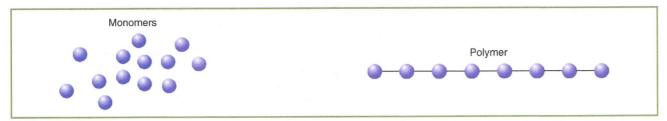

Figure 1.2 Monomers and polymers.

There are many natural examples of polymers: rubber, proteins, DNA, starch, etc. Synthetic polymers include polystyrene (used to make CD jewel cases) and PVC (polyvinyl chloride, used in plastic wrap and water pipes). An elastomer has cross-linked polymer chains. In Silly Putty, boric acid causes cross-linkage between different silicone polymers.

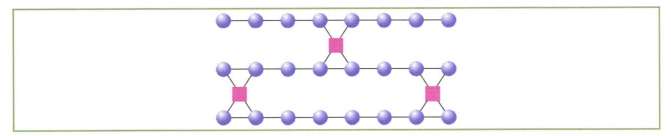

Figure 1.3 An elastomer.

Making real Silly Putty is expensive and requires special safety precautions. A good substitute is commonly known as gluep. Polyvinyl acetate (school glue, aka Elmer's Glue) is used in place of silicone polymer. The cross-linking agent is the same: borax solution. Borax is a naturally occurring mineral that is used as a cleaning agent. It may be irritating to the eyes, so care should be used in handling it; wash your hands when you are done.

Your job today is to figure out the best recipe for making a Silly Putty-like elastomer. You should take an experimental approach. Based on what you know, develop a hypothesis about what the relative proportions should be. As you think about designing your experiment, consider the independent and dependent variables, control group and standardized variables. (Hint: Handling the product is helpful.) Each group will make two different experimental elastomers.

The materials at your disposal are:

1. School glue (up to 30mL)
2. Borax solution (5%) (up to 30mL)
3. Water (up to 30mL)
4. Beaker
5. Stirring rod
6. Scoopula

The glue and water should be thoroughly mixed together before the cross-linking agent is added.

Answer the follwing question.

What is your hypothesis?_____
_____

What is your prediction?_____
_____

What is the independent variable?_____
_____

What is the dependent variable?_____
_____

How will you measure it?_____
_____

Describe, in broad terms, your method. _____
_____
_____

Did your experiments fully test your hypothesis?_____
_____

What new questions do you have? What would you do differently?_____
_____
_____
_____

COMMUNICATING SCIENCE

Scientists don't work in isolation. Individual scientists make up a science community; an individual scientist is not part of the community unless and until he or she communicates what he or she has learned. Initially, communication takes place informally, in hallways, journal clubs and scientific meetings. But the eventual goal is publication in a peer-reviewed journal.

The peer-review process is intended to vet the contribution. Is the question important, the approach technically correct, the data properly analyzed, the results appropriately interpreted? Peer review is a group process. A researcher sends a manuscript to an editor of a journal. The editor sends the manuscript to several experts in the field. These reviewers critique the manuscript and recommend whether to publish. A good editor will "review the reviews" in making the final decision. Acting as an editor or reviewer is part of service to the scientific community; it is unpaid work.

Articles may be published in journals ranging from the most prestigious, international journals such as *Science* or *Nature*, to local journals such as *Virginia Journal of Science* or special topics such as *Journal of College Biology Teaching*. Many, if not most, journals are available online, some, such as *PLOS (Public Library of Science) Biology* with open access.

Does the peer-review system work? Although the process is not perfect, it seems to work fairly well. In a recent egregious case, South Korean stem-cell researcher Hwang Woo-Suk was shown to have fabricated data. However, scientists are concerned that with the recent explosion in publications, the limits of the system are being approached. As always with science, new ideas are constantly being tested. For example, the journal *Nature* has just conducted a trial of open peer-review (posting submitted

> Ultimately, worth is assessed by whether the scientific community decides to build on a particular finding.
>
> Parthasarathy H. (2005). Published and not perished. Public Library of Science Biology 3(10), p. e367. doi:10.1371/journal.pbio.0030367

manuscripts online for open comment). It was not successful, in large part due to poor participation. (See www.nature.com/nature/peerreview/debate/index.html for more info.)

Usually, only professionals (or budding scientists) read the peer-reviewed or "primary" literature. In the hierarchy of science reporting, magazines that synthesize recent science news and present it in a form that is accessible to a general audience are written by scientists or professional science writers. Some examples include *Scientific American, Science News, New Scientist, Discover, National Geographic,* and *Popular Science*. Usually such articles are trustworthy sources of information.

Yet with unlimited presentation of news, via television and internet, disinformation is common. One reason is a misguided effort to be "fair and balanced" in science reporting. This is a grave problem.

"Fair and balanced," however, doesn't mean putting all viewpoints, regardless of their underlying logic or validity, on an equal footing. Discerning the merits of competing claims is where the empirical basis of science should play a role. I cannot stress often enough that what science is all about is not proving things to be true but proving them to be false. What fails the test of empirical reality, as determined by observation and experiment, gets thrown out like yesterday's newspaper. One doesn't need to debate about whether the earth is flat or 6,000 years old. These claims can safely be discarded, and have been, by the scientific method. (Lawrence M. Krauss, *Scientific American*, December 2009)

Why is this of general interest? Because science is a part—a large part—of our culture. In our day-to-day lives we make decisions based on the information we obtain. These decisions can impact public policy, including policy regarding research, and the public good.

A recent example is summarized in a January/February 2010 *Discover* magazine article by Andrew Grant: "*Vaccine phobia becomes public-health threat.*" There's been a significant rise in cases of autism over the last several decades. Initial research suggested a connection with the mercury used in childhood vaccinations. Subsequent research, over and over, has shown no relationship. In addition, mercury in vaccinations has been decreased to trace amounts. And a genetic basis of autism has been well-documented in some cases.

Nonetheless, fearful parents have declined to have their children vaccinated and have been vocal in their anti-vaccination stance. Several negative consequences can result. First, outbreaks of vaccine-preventable childhood illnesses such as measles have already occurred. Measles and other childhood diseases are potentially life-threatening. Second, such outbreaks have the potential to become public health issues. If the disease agents (e.g., viruses) are active and circulating in the population, the possibility of evolution to more virulent, resistant strains exists. If that were to occur, even vaccinated children would be at risk. Third, if funding for autism research continues to be directed to the discredited autism-vaccine link, progress toward truly understanding and preventing or treating autism will be delayed.

Science News Article Assignment

Read a science news article. Article criteria include: topic in biology, reputable source (major newspaper or lay science journal, either print or on-line), at least 500 words, published within the last 3 months).

Read, summarize and review a science news article. Your report should include the following:

Title of Article

Author

Source

Original Date of Publication

Summary (Summarize the article in 1–3 paragraphs.)

Analysis (Analyze the article. Is the biology clearly explained? Has the author made a justifiable interpretation of the data? Is the original source of the info cited or referenced in any way?)

Questions (List 5 questions that came to mind.) e.g., the meaning of a term, or additional info you'd like to have. Find out and report the answer to at least one.)

Public Health/Social Significance (What public policy issues does it concern? Is it important in your life, or the life of someone you know?)

You must provide a copy of the full article including all relevant bibliographic information.

LAB 2

Measurement: Metrics and pH

The United States is one of the few countries in the world that commonly uses the English system of measurement rather than the metric system. Scientists across the world use the metric system because it is simpler (based on units of 10), which makes it easier to convert between units of measurement and to compare data across experiments. All conversions within the metric system are done by moving the decimal point to the right or the left, which is the same as multiplying or dividing by factors of 10.

Common Terminology in the Metric System: (see conversion chart on next page)

BASE UNITS OF MEASUREMENT
    Distance—meter (size of a meter stick)
    Volume—liter (l)
    Weight—grams (g)

Prefix for LARGE measurements
    Kilo (k) = thousand (10^3; e.g., a kilometer is 1,000 meters)

Prefixes for SMALL measurements
    Centi (c) = hundredth (10^{-2}; e.g., a centimeter is 1/ 100th of a meter)
    Milli (m) = thousandth (10^{-3}; e.g., a millimeter is 1/ 1,000th of a meter)
    Micro (u) = millionth (10^{-6}; e.g., a micrometer is 1/ 1,000,000th of a meter)
    Nano (n) = billionth (10^{-9}; e.g., a nanometer is 1/ 1,000,000,000th of a meter)

We can use these prefixes (e.g., **kilo, centi**) before any type of measurement (**kilogram, milliliter**) to specify a particular measurement. For example, if you are measuring really long distances, you would use **kilometers** (km), which equals 1000 meters; if you are measuring small distances (like your fingernail) you would use **centimeters** (cm)or **millimeters** (mm) that are found on a meter stick or ruler. The same is true when measuring a large weight, **kilograms** (kg), or a small volume, **milliliters** (ml).

Converting within the Metric System

Now let's discuss converting within the metric system. Very often, when using the metric measurements, conversions need to be made from a larger unit of measurement (e.g., a meter) to smaller unit of measurement (e.g., centimeters or millimeters), or vice versa. Fortunately, this is easy to do in the metric system, because it is based on multiplying or dividing units of 10. This means you only need to move the decimal point to the right or left depending on the direction the conversion is happening.

Adapted from *Laboratory Manual for the Processes of Life: BIO 101* by Holyoke Community College. Copyright © 2013 by Kendall Hunt Publishing Company. Reprinted by permission.

Converting from Large to Small

If you are converting from a LARGER measurement to a SMALLER measurement, the decimal will move to the RIGHT (**LSR**) because you are multiplying by degrees of 10 (look at the conversion chart and if it is higher in the list, it is larger). For example, if you are measuring distance, you will use the meter (m) as the base. If you are measuring a long distance, you will normally see this expressed in *kilo*meters (km).

2 km = _____m : 1 km = 1000 m or 3 decimal places (10^3), so 2 km = 2.0, and then move the decimal three places to the right (large to small); 20.0, 200.0, 2000.0. So 2 km = 2000 m

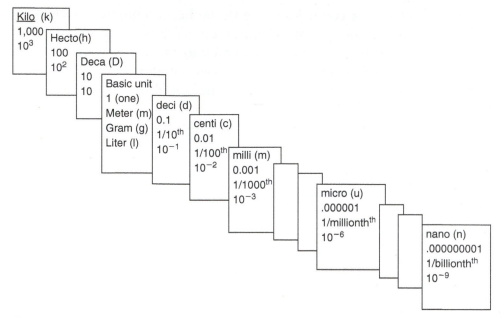

Conversion Chart

Practice

1. 8 km = _____m
2. 3.5 km = _____m

If you are converting from a SMALLER measurement to a LARGER measurement, the decimal will move to the LEFT (**SLL**). For example, if you are measuring distance, you will use the meter (m) as the base. If you are measuring a short distance, you will normally see this expressed in *centi*meters (cm).t

2 cm = _____m : 1 cm = .01 m or 2 decimal places (10^{-2}), so 2 cm = 2.0, and you move the decimal two places to the left; 0.2; 0.02, so 2 cm = 0.02 m

Practice

3. 9 cm = _____m
4. 3.5 cm = _____m

If you are converting between different units of measurements, first decide whether you are converting from large -> small or small -> large; remember, large -> small = right (**LSR**) and small -> large = left (**SLL**). Then **use the conversion chart to figure out how many spaces you will be moving the decimal.** For example if you are converting 2 cm to _____nm, use the **LSR** (large -> small = right) rule, and then use the conversion chart to determine how many spaces between cm and nm (there are seven). Using the LSR rule, you will be moving the decimal point seven places to the right. So 2 cm = 20,000,000 nm.

Likewise, if you are converting 200 um to ___mm, use the **SLL** (small -> large = left) rule and then use the conversion chart to determine that there are three spaces between um and mm; you'll be moving the decimal point three spaces to the left: 200 um = .2 mm.

Distance Exercises

- Looking at the *meter stick*, it is broken into 100 portions. How long is the distance from the "0" end of the meter stick to the number 1, or how long is each of the 100 portions? _____
- How many hash marks are there between each number on the meter stick? _____What is this measurement?_____
- Looking at the *ruler*, how long is it in centimeters? _____ How long is the ruler in millimeters? _____

- Complete the following measurements:
   a. Distance from the floor to the top of the table _____m; _____cm; _____mm; _____km
   b. Distance from the floor to the top of the light switch _____m; _____cm; _____mm; _____km
- Knowing that bacterial cells are up to 100 times smaller than one of your own cells, what units would you use to measure their diameter? _____ Why?

MEASURING VOLUME

Volume is the three-dimensional space that an object takes up. You are probably most familiar with using volume to measure an amount of liquid. A liter is the base unit when measuring liquid volume. However, if you want the volume of a solid rectangular or square object, you can use cm^3 ($1\ cm^3 = 1\ ml$), which is calculated by:

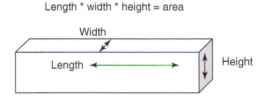

Length * width * height = area

Volume Exercises

1. Get a graduated cylinder, a large and small test tube, a wooden block, and a metal slug.
2. What is the stated volume that is measured by the graduated cylinder? _____
3. What is the volume of each type of test tube?
   a. Large: _____ ml; _____ l (Hint: You can simply convert from one measurement to the other. You don't have to do each one separately.)
   b. Small: _____ ml; _____ l.
4. What is the volume of the block? _____.
   a. What is the identification number of your block? _____
   b. Which unit did you use as your base? _____ Why?

5. Explain how you could determine the volume of your metal slug. [Hint: You can use your graduated cylinder to help figure out the volume of the metal slug.]

6. What is the volume of the metal slug?_____
   a. Which unit did you use as your base? _____Why?
   b. What would the volume be in liters? _____
   c. What is the metal slug identification number? _____

MEASURING WEIGHT

Mass is the measure of the amount of matter in an object and weight is the measure of the pull of gravity acting on the mass. (Both mass and weight are measured in grams.)

Weight Exercises

- What is the weight of the metal slug? _____ g; _____ mg; _____ kg
- What is the weight of the wooden block? _____ g; _____ mg; _____ kg
- What is the weight of a large test tube? _____ g; _____ mg; _____ kg
- What is the weight of a small test tube? _____ g; _____ mg; _____ kg

MEASURING TEMPERATURE

Temperature is measured using the Celsius scale. $°C = (°F − 32) * 5/9$. Note that in Celsius that freezing of water is at 0 and that boiling is at $100 °C$.

Temperature Exercises

1. Get a beaker and a thermometer.
2. What is the temperature of the room?_____
3. What is the temperature of the water in the ice bath?_____
4. Now hold the thermometer in your hand for 5 minutes. What is the temperature of your skin?_____
   a. Is this the same as a "normal" person's body temperature of $37.0 °C$? _____ Why or why not? _____

MEASURING pH

pH measures the amount of hydrogen ions (H^+) there are in a solution. The pH scale ranges from 0–14: below 7 is considered acidic, 7 is considered neutral, and above 7 is considered basic. The pH scale is logarithmic, meaning a change of 1 unit means the liquid is 10 times more acidic or basic (change of 2 units is 100 times more acidic or basic). To measure pH you can use a pH meter, pH paper, or a chemical that will change color based on the pH of the solution.

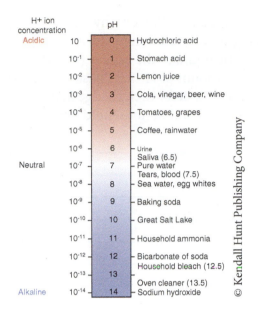

pH Exercises

1. Phenol red is a chemical that is red in the presence of a liquid that has a pH => 8, colorless when the solution is between 8 and 6.4, and yellow when the solution has a pH <=6.3.

pH	1	2	3	4	5	6	7	8	9	10	11	12	13	14
Color		Yellow				>-⌐<-	colorless				Pink to Red			

   a. What pH range do you expect distilled water to have? _____ If your hypothesis is correct, what color should result when phenol red is added to the distilled water? _____

   b. Get a beaker of distilled water (25 ml) and add 5 drops of phenol red (make sure to swirl to mix the liquids). What color is the solution? _____ Does it match the color you expected?_____ What can you conclude about the pH of our distilled water?

   c. NaOH produces many OH^- ions that will bond with H^+ thus removing them from the solution. If you add NaOH to your distilled water/phenol red solution, what change in color do you expect? _____ Why?

   d. Record the number of drops of NaOH it takes to see a color change (swirl the beaker after every other drop): ___ What is the color change?_____. Is this the color you hypothesized? _____.

   e. What is the pH of the liquid in the beaker (use pH paper to determine it, by dropping a little liquid onto the paper strip and after waiting for a minute, match to the code on the container)?_____

   f. If you know that $CO_2 + H_2O \rightarrow H_2CO_3 <-> HCO_3^- + H^+$, what should happen to the color of the solution as you add CO_2?

   g. How can we easily add CO_2 to the solution (Hint: There are straws available.)?

   h. Now perform this and watch until you see the color change. What color is the solution now? _____

    i. Is this the color that you expected? Why or why not?

    j. What should the pH range of the solution be now?_____

    k. What is the actual pH of the liquid in the beaker (use pH paper to determine it)?_____

    l. What is the change in the amount of H^+ ions based on the starting pH (answer "e") and what you just measured (remember that for each change in pH there is a tenfold increase in H^+ ions)?

2. Use pH paper to examine some common household items:

Item	Hypothesis	Actual
Vinegar		
Apple Juice		
Orange Juice		
Milk		

LAB 3

Organic Chemistry

Living organisms are made of four major organic compounds: carbohydrates, proteins, lipids, and nucleic acids. Each of these macromolecules is held together by covalent bonds and has different structures and properties. For example, lipids (usually made of fatty acids and other subunits) have many C–H bonds and relatively little oxygen, while proteins (made of amino acids) have amino groups (–NH2) and carboxyl (–COOH) groups. These characteristic subunits and groups impact the chemical properties of macromolecules. For example, monosaccharides such as glucose are polar and soluble in water, whereas lipids are nonpolar and insoluble in water.

Functional group	Structural formula	3-D Model	Functional group of:
Hydroxl	–OH	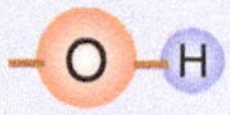	Alcohols, carbohydrates
Sulfhydryl	–S–H	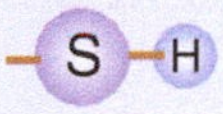	Proteins, rubber
Carbonyl			Formaldehyde
Carboxyl		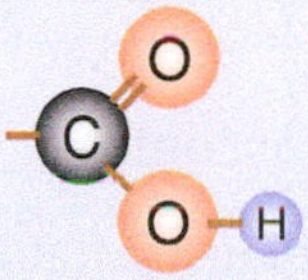	Amino acids, vinegar
Amino		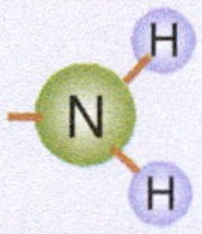	Ammonia
Methyl		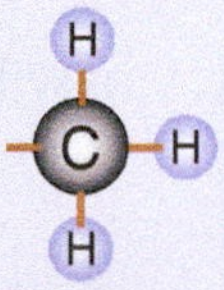	Methane
Phosphate		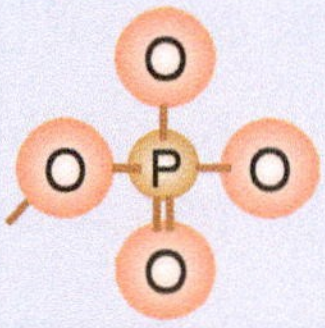	ATP, nucleic acids, phospholipids

Figure 3.1 Functional Groups

© Kendall Hunt Publishing Company

CONTROLLED EXPERIMENTS CAN IDENTIFY ORGANIC COMPOUNDS

Scientists have devised several biochemical tests to identify the major types of organic compounds in living organisms. Each of these tests involves **experimental samples** to be identified and **controls**. Experimental samples may or may not contain the substance the investigator is trying to detect. Only a carefully conducted experiment will reveal its contents. In contrast, **controls are known solutions**. We use controls to validate that our procedure is detecting what we expect it to detect and nothing more.

A positive control contains the variable for which you are testing; it reacts positively and demonstrates the test's ability to detect what you expect. For example, if you are testing for protein in unknown solutions, then an appropriate positive control is a solution known to contain protein. A positive reaction shows that your test reacts correctly; it also shows you what a positive test looks like.

A negative control does not contain the variable for which you are searching. It contains only the solvent, often distilled water, with no solute and will not react in the test. A negative control shows you what a negative result looks like. If a negative control reacts to a biochemical test, it demonstrates a problem with the experiment such as a mislabeled solution or contamination.

Controls are also important because they reveal the specificity of a particular test. For example, if water and a glucose solution react similarly in a particular test, the test cannot distinguish water from glucose. If the glucose solution reacts differently from distilled water, the test can distinguish water from glucose. In this instance, the distilled water is a negative control for the test and a known glucose solution is a positive control.

It is important to identify controls before beginning an experiment.

In the lipid grease spot test, you will use vegetable oil, distilled water, and an unknown.

1. Which of these solutions is the positive control? _____
2. Which of these solutions is the negative control? _____

In Benedict's test of reducing sugars (monosaccharides), you will use potato juice, sucrose, glucose, distilled water, fructose, an unknown, and starch.

3. Which two of these solutions can be used as positive controls?

_____   _____

4. Which of these solutions is the negative control? _____

In the iodine/Lugol's test for starch, you will use potato juice, sucrose, glucose, distilled water, fructose, an unknown, and starch.

5. Which of these solutions is the positive control? _____
6. Which of these solutions is the negative control? _____

In Biuret's test for proteins (peptide bonds), you will use egg albumin, amino acids, distilled water, an unknown, and protein.

7. Which of these solutions is the positive control? _____
8. Which of these solutions is the negative control? _____

As you begin the actual experiment, remember to think critically as you record your results. If you question something, discuss it with your lab partners and your instructor.

Lipids

Lipids include a variety of molecules that dissolve in nonpolar solvents such as ether and acetone, but not in polar solvents such as water. Triglycerides are abundant lipids that are made of glycerol and three fatty acids. The grease spot test for lipids is a simple test based on the ability of lipids to produce translucent grease marks on unglazed paper.

PERFORM THE GREASE-SPOT TEST FOR LIPIDS

1. Using pencil marks, divide one square of unglazed brown paper into four quadrants. Label each quadrant 1 to 4 or write the name of the substance that will be added to that area.
2. Add one small smear of vegetable oil to quadrant 1. When applying solutions to the paper, do not allow the solutions to bleed into each other. If this occurs, you will need to start again.
3. Add one small smear of distilled water to quadrant 2.
4. Obtain a tube of unknown solution and a pipette. Record your unknown number on <u>each and every</u> data table in this chapter.
5. Add one small smear of unknown to quadrant 3. Quadrant 4 will be empty.
6. Place the paper in a dry place on the desk top.
7. Observe for changes in one hour, after the negative control is dry.
8. Record observations in the table below. Describe each quadrant as dry, crusty, or translucent.
9. Throw all used tape, brown paper, and pipettes in the trash.

TABLE 3.1 Grease Spot Test for Lipids				
Corner	Solution	Description	Positive	Negative
1	1 drop Salad Oil			
2	1 drop Distilled H_2O			
3	1 drop Unknown Solution #_____			

Carbohydrates

Carbohydrates are molecules made of carbon (C), hydrogen (H), and oxygen (O) in a ratio of 1:2:1. The chemical formula for glucose is $C_6H_{12}O_6$. Glucose is an example of a monosaccharide or a reducing sugar. Paired monosaccharides form disaccharides. Sucrose (table sugar) is a disaccharide linking glucose to fructose. Similarly, linking three or more monosaccharides forms a polysaccharide. Starch, glycogen, chitin, and cellulose are examples of polysaccharides.

Monosaccharides are reducing sugars. Benedict's reagent, a blue solution, is used to indicate the presence of reducing sugars. The presence of reducing sugars is indicated when the solution, after heating, changes colors. A green solution indicates a small amount of reducing sugars and a reddish orange or a brown (rusty) color indicates an abundance of reducing sugars. All of these colors indicate the presence of monosaccharides and signify a positive test for reducing sugars. Non-reducing sugars and non-sugars, such as sucrose or albumin, will remain blue during Benedict's test, indicating a negative result.

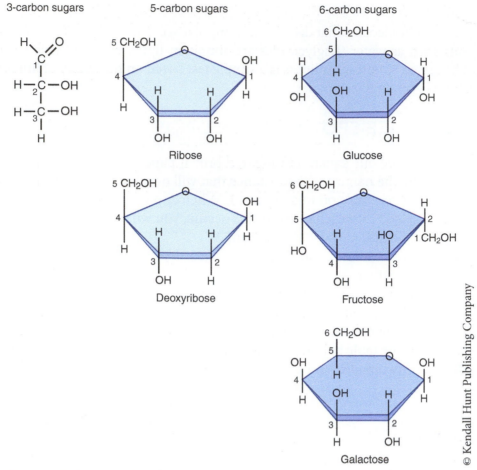

Figure 3.2 Monosaccharides

PERFORM BENEDICT'S TEST FOR REDUCING SUGARS (MONOSACCHARIDES)

1. Add approximately 200 mL of distilled water to a large beaker and place the beaker on the hot plate.
2. Turn the hot plate on high or to 300°C and allow the distilled water to come to a boil.
3. Using masking tape, label seven **large test tubes** with numbers 1 to 7. Place the tape near the top of the test tube because you will immerse the test tube in boiling distilled water.
4. Add 2 ml of Benedict's reagent** to each of the labeled test tubes.
5. Add 10 drops of each solution listed in the data table (Table 2) to the correctly labeled test tubes.
6. Carefully place all seven test tubes into the boiling water bath. Boil the test tubes for 3 minutes.
7. After three minutes have elapsed, carefully remove test tubes from boiling water bath using the test tube holder. Place hot tubes in the test tube rack and allow the tubes to cool.
8. Gently swirl the tubes, and observe for color changes.
9. Record observations in Table 2. Describe the color the solution is at the end of procedure and place a check mark under the positive or negative column as appropriate.
10. Pour waste solutions into the labeled hazardous waste containers. Scrub and dry test tubes. Place all used tape and pipettes into the trash.

Special Note*: **Benedict's reagent is a moderately toxic skin and eye irritant. Goggles must be worn when handling this reagent. It is harmful to the environment. It is a **hazardous waste.**

TABLE 3.2 Benedict's Test for Reducing Sugars (Monosaccharides)					
Test Tube	Solution	Reagent	Color Reaction	Positive	Negative
1	10 Drops potato juice	2 ml Benedict's			
2	10 drops sucrose solution	2 ml Benedict's			
3	10 drops glucose solution	2 ml Benedict's			
4	10 drops distilled H_2O	2 ml Benedict's			
5	10 drops fructose Solution	2 ml Benedict's			
6	10 drops starch Solution	2 ml Benedict's			
7	10 drops unknown solution #____	2 ml Benedict's			

Starch

Staining by Lugol's solution (I_2KI), a yellowish brown fluid, distinguishes starch from other carbohydrates. Iodine interacts with the coiled molecules found in starch and becomes bluish black or greenish black in color. Iodine does not react with carbohydrates that are not coiled and will remain yellowish brown. Therefore, a bluish or greenish black color is a positive test for starch. Any other change in color or no change in color is a negative test for starch.

PERFORM THE LUGOL'S TEST FOR STARCH

1. Using masking tape, label seven small test tubes with numbers 1 to 7.
2. Add 10 drops of each solution listed in Table 3 to the correctly labeled test tubes.
3. Add 2 drops of Lugol's** solution to each test tube.
4. Gently swirl the tubes, and observe for color changes.
5. Record observations on data sheet. Describe the color the solution is at the end of procedure and place a check mark under the positive or negative column as appropriate.
6. Dispose of tube solutions down the drain. Scrub and dry test tubes. Dispose of all used tape in the trash.

TABLE 3.3 Iodine* test for Starch * LUGOL'S SOLUTION = IODINE POTASSIUM IODIDE (I_2KI)					
Test Tube	Solution	Reagent	Color Reaction	Positive	Negative
1	10 Drops potato juice	2 drops Lugol's			
2	10 drops sucrose solution	2 drops Lugol's			
3	10 drops glucose solution	2 drops Lugol's			
4	10 drops distilled H_2O	2 drops Lugol's			
5	10 drops fructose Solution	2 drops Lugol's			
6	10 drops starch Solution	2 drops Lugol's			
7	10 drops unknown solution #____	2 drops Lugol's			

Special Note: **Lugol's is irritating to eyes and skin. Goggles must be worn when handling or using this solution. It is not toxic to the environment and may be disposed of by pouring into the sink drain and flushing with water.

Proteins

Proteins are remarkably versatile structural molecules found in all life forms. Proteins are made of amino acids, each of which has an amino group (NH2) and a carboxyl acid group (COOH). The bond between these two groups found on adjacent amino acids in a protein is a peptide bond, which is a type of covalent bond.

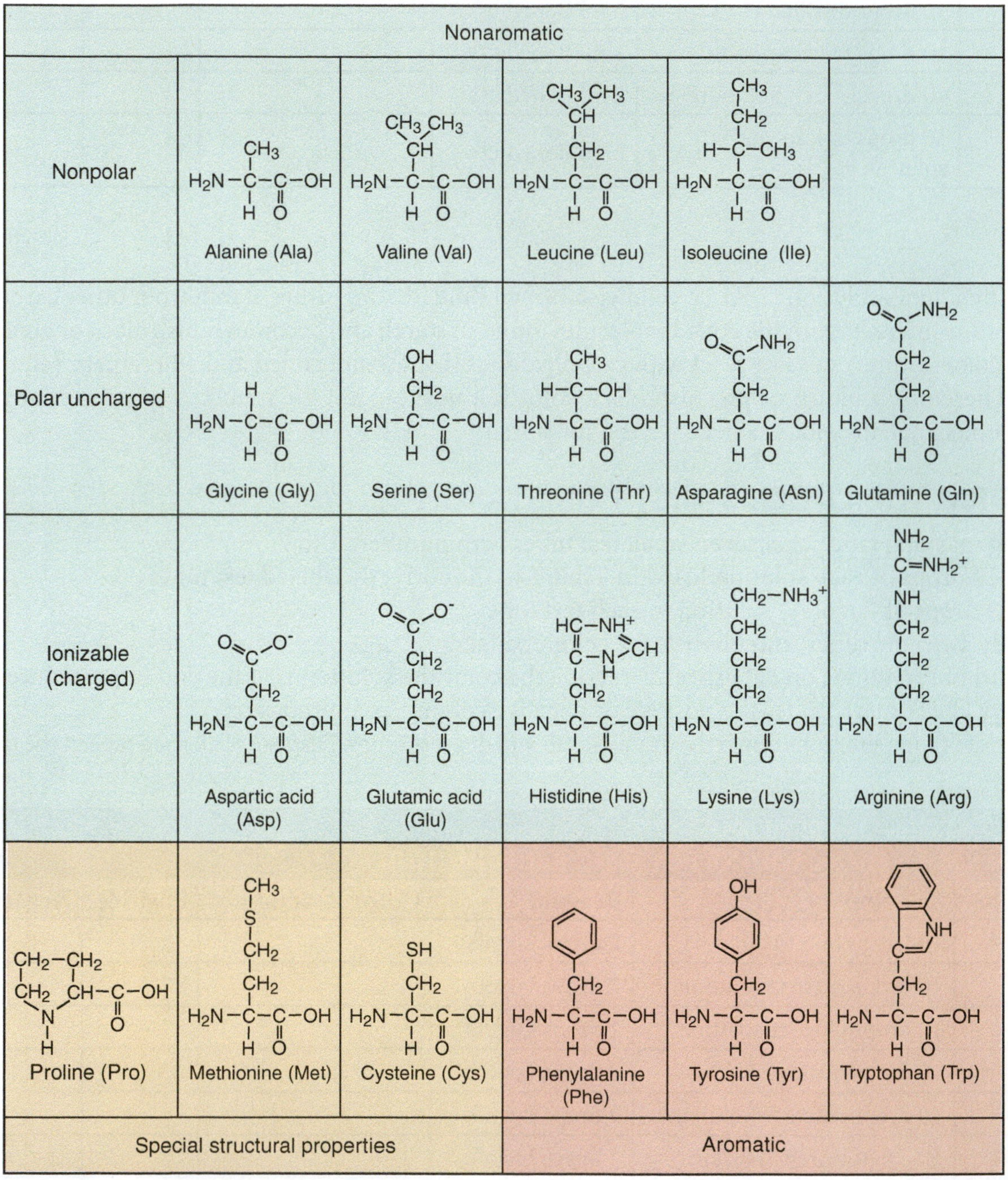

Figure 3.3 Amino Acids

Biuret's reagent is used to identify peptide bonds. The copper sulfate in Biuret's reagent gives the reagent a blue color. A violet color is a positive test for the presence of peptide bonds, which proteins contain in large numbers. If the solution remains blue, it is negative for the presence of peptide bonds.

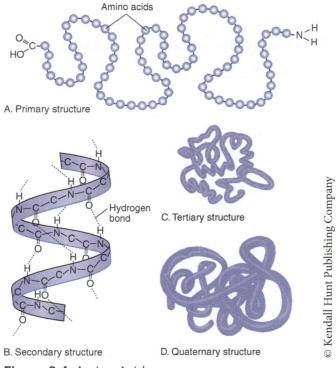

A. Primary structure

B. Secondary structure

C. Tertiary structure

D. Quaternary structure

Hydrogen bond

Amino acids

© Kendall Hunt Publishing Company

Figure 3.4 Amino Acids

PERFORM THE BIURET'S TEST FOR PROTEIN

1. Using masking tape, label five large test tubes with numbers 1 to 5.
2. Add 2 ml of Biuret's reagent to each labeled test tube.
3. Add 2 ml of each solution in Table 4 to the correctly labeled test tubes.
4. Gently swirl the tubes, and observe for color changes.
5. Record observations on data sheet. Describe the color the solution is at the end of procedure and place a check mark under the positive or negative column as appropriate.
6. Dispose of tube solutions in the marked hazardous waste containers. Scrub and dry the test tubes. Throw used tape and pipettes in the trash.
7. Do free amino acids have peptide bonds? _____

TABLE 3.4 Biuret's Test for Proteins with Peptide Bonds					
Test Tube	Solution	Reagent	Color Reaction	Positive	Negative
1	2ml egg albumin solution	2 ml Biuret's			
2	2ml amino acid solution	2 ml Biuret's			
3	2ml distilled H_2O	2 ml Biuret's			
4	2 ml protein solution	2 ml Biuret's			
5	2 ml unknown solution #____	2 ml Biuret's			

Special Note: **Biuret's reagent is harmful to humans and the environment. It is a corrosive liquid; especially dangerous to eyes!!** You must wear goggles when in the presence of this reagent. It is a concentrated sodium hydroxide (308 grams in 770 ml of distilled water) and copper sulfate pentahydrate solution. It is a **hazardous waste**.

IDENTIFY ORGANIC COMPOUNDS IN YOUR UNKNOWN

1. Using data you collected about your unknown, circle the organic compounds your unknown contains:

   Number of "unknown solution"_____

   a. Monosaccharide or reducing sugars
   b. Starch
   c. Protein with a peptide bond
   d. Lipid
   e. None of the above

LAB 4

Scope and Cells

Biological molecules come together in complex ways to form cells—the fundamental unit of life. The sizes, shapes, and functions of cells are myriad. A good way to take a quick survey of cells in the various kingdoms is to use the compound light microscope.

THE MICROSCOPE

The compound light microscope (aka scope) is a mechanical device. It does not have any digital components. You must manipulate this machine, using the appropriate parts, to make it work for you. Use the scope carefully but don't be afraid to use the moving parts.

How Tos

How to Carry the Scope

1. Always use both hands.
2. Carry with one hand under the base and the other in the recessed handle if available.
3. Never slide the scope across a surface; pick it up to reposition it.

How to Set up the Scope

1. Set the scope gently on bench top.
2. Remove the dust cover; fold and put it out of the way.
3. Unwrap the cord from body of scope; plug into outlet.
4. Stage should be positioned at the midpoint of its range.
5. Lowest (shortest, red-lined, 4×) objective lens should be in position.
   a. If not, use the revolving nosepiece to position it; it will click into place.

How to Get Started

1. Turn on the light. Note that the light comes from beneath the stage.
2. Turn the rheostat (dimmer switch); observe the change in light intensity.

How to Adjust the Scope for Your Face (Interpupillary Distance)

1. Look through the ocular lenses. (No specimen necessary.)
2. Use the rheostat to adjust the light intensity for your comfort.
3. Hold the ocular lenses with both hands and adjust their position (as you would a pair of binoculars) until you see one circle of light (see figure below, left).
   a. This may not be entirely comfortable, but will become more so with practice.
   b. If you look through just one ocular, you will experience eyestrain.

Adapted from From *Biology 101 Lab Manual* by Ann S. Evans, Sarah Finch, Eric Lamberton and Randy L. Durren. Copyright © 2014 by Kendall Hunt Publishing Company. Reprinted by permission.

4. If the field of view (the circle you see) is only partially filled with light, slowly adjust the field iris diaphragm by moving the ring until the field of view is filled (see figure below, right).

   a. At higher magnifications you will readjust the field iris diaphragm.

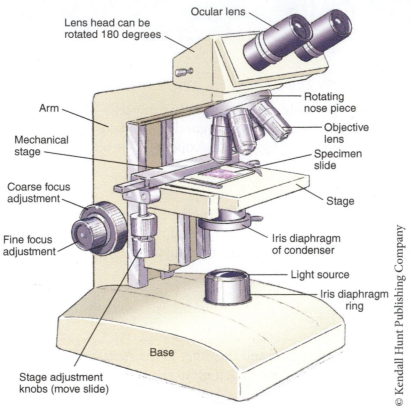

Ocular lens

Lens head can be rotated 180 degrees

Arm

Mechanical stage

Coarse focus adjustment

Fine focus adjustment

Stage adjustment knobs (move slide)

Base

Rotating nose piece

Objective lens

Specimen slide

Stage

Iris diaphragm of condenser

Light source

Iris diaphragm ring

© Kendall Hunt Publishing Company

Figure 4.1 Parts of the compound microscope.

CARE AND SAFETY

1. Use only lens paper to clean the lenses. Never use paper towel or tissue. Ask your instructor, who has lens paper and cleaning fluid, to assist you.
2. Always use a cover slip. DO NOT allow water and the salt or dye it contains to touch the lens or the stage. You may carefully blot a small spill on the stage dry, using a small piece of paper towel. If liquid is near the objective lens, ask your instructor for assistance *immediately*. Otherwise the lenses may be permanently damaged.
3. Minor scratches or even cracks in the cover slip will not impede your work; you will focus past them.

How to Put a Slide on the Stage

1. Make sure the stage is low enough to accommodate placing the slide.
2. Make sure the lowest objective is in place.
3. Open the specimen holder by moving the curved finger knob.

4. Place the slide on the stage and slide it back as far as it will go to the lower left.
5. Gently release the curved finger knob.
6. DO NOT put the curved finger on top of the slide.

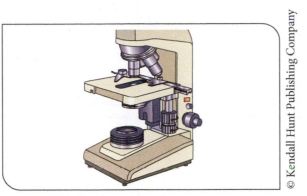

Figure 4.2 Correct slide placement
© Kendall Hunt Publishing Company

How to Move the Specimen

1. Move the mechanical stage adjustment knobs.
2. DO NOT move the stage with your hand.
3. Position the slide so that the specimen is centered over the light. If in doubt, start by centering on the middle of the cover slip.

HOW TO FOCUS ON A SPECIMEN

1. Always start with the lowest (4×) objective engaged.
2. Insert the slide as above.
3. Using the coarse focus knob, position the stage as close as possible to the objective lens.
4. While looking through the oculars, use the coarse focus knob to slowly move the stage toward the objective lens until you can see the specimen.
5. To make sure you have the best possible focus, continue using the coarse focus until you go past the best focus and then back to the best.
6. Use the fine focus knob in the same way to get the best possible view.
7. Position the specimen in the center of the field of view.
8. Move to the 10× objective (you should be able to hear it click into place).
   a. Do not move the stage or the position of the slide on the stage at this point.
9. Use the coarse focus and then fine focus, as before.
10. Next move to the highest (40×) objective.
    a. DO NOT use the coarse focus knob.
    b. Use only the fine focus knob at the highest objective.
    c. Adjust the light using:
       i. aperture iris diaphragm by moving the lever.
       ii. field iris diaphragm by moving the ring.

TABLE 4.1 Comparison Among Objectives

Objective	4×	10×	40×
Total magnification	40×		
Diameter of field of view	5 mm (=5000 μ)	2 mm (=2000 μ)	
Working distance	4x ↕ 25 mm	10x ↕ 8.3 mm	40x ↕ 0.5 mm
Depth of view	175 μ	28 μ	3 μ

© Kendall Hunt Publishing Company

Important Concepts

Magnification Is Multiplicative

When you are using the 4× objective lens, the image is magnified four times. Then the ocular lens magnifies that image 10 times, for a total of 40×. Fill in the total magnification for each objective lens in the table.

The Diameter of the Field of View Decreases with Increasing Magnification

At 40×, the diameter of the field of view is 5 mm (see Table 1). At 100×, the diameter of the field of view is 2 mm. At 400×, which is a tenfold greater magnification than at 40×, the diameter of the field of view is _____

_____

The Working Distance Decreases with Increasing Magnification

In other words, there is more space between the objective lens and the specimen at 4× than 10×, and less space the higher the objective lens. At 400×, the space is only 0.5 mm (see Table 1). Why must we **not** use the coarse focus knob at total magnification of 400×?_____

_____

The Depth of View Decreases with Increasing Magnification

The distance through which you can focus is 175 μ at 40× total magnification, but only 3 μ at 400× total magnification. If you want to focus through an object, which magnification should you use?

_____

The Objectives Are Parfocal

When you move to the next highest objective, the specimen should already be focused enough to be discernable. You should have to do minimal focusing to get the image in focus with the new objective. Once you have focused well at 40×, should you move the stage down before going to 100× ? _____ What about going from 100× to 400×?

_____

The Objectives Are Parcentric

This is why you should center the specimen in the center of the field of view. When you move to the next highest objective, it is the center of the field of view that is further magnified. If the specimen of interest is at the edge of the field of view, it will not be visible when you increase magnification. If you "lose" the specimen, which would be the best approach: start over at the lower magnification or search at the higher magnification?

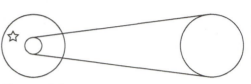
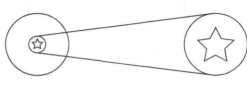

Figure 4.3

_____

PRACTICE USING THE SCOPE

Two prepared slides, "Letter e" and "Colored Threads," can both be used to practice focusing on an object. Some points described above can be illustrated with either slide. Different additional points can be made with "Letter e" and "Colored Threads."

Because the diameter of the field of view decreases as the magnification increases, the amount of the specimen that you can see also decreases.

At higher magnification, as the working distance decreases, the image contrast will also decrease. You can improve the image by adjusting the light, using the iris diaphragm lever or the field diaphragm ring. Be aware that contrast (perceivable difference between two objects) and resolution (ability to discern detail) cannot be simultaneously maximized. Use the light and fine focus to obtain the best image for your purposes.

Colored Threads Slide

1. Hold the slide up to the light. Notice that three threads (blue, red, and yellow) are visible with the naked eye.
2. Using the "How to focus on a specimen" procedure, focus on the specimen at 40× total magnification.
    a. Remember to use the stage adjustment knobs to center the specimen.
    b. Can you bring all three threads into focus at the same time? _____
    c. Focus through the threads using the fine focus.
        i. Can you tell which thread *appears* to be on top? _____
    d. In Figure 4.4, sketch your field of view. "To scale" means draw it as you see it. If the specimen takes up a third of the diameter of the field of view, draw it that way.
3. Focus at 100× total magnification:
    a. Can you see all three threads? _____
    b. Can you see more or less of the specimen than at 40×? _____
    c. In Figure 4.4, sketch your field of view to scale.
4. Focus at 400× total magnification:
    a. Can you see all three threads? _____
    b. Can you bring all three threads into focus at the same time? _____
    c. Can you focus through as much as you could at 100× or 40×? _____
    d. In Figure 4.4, sketch your field of view to scale.

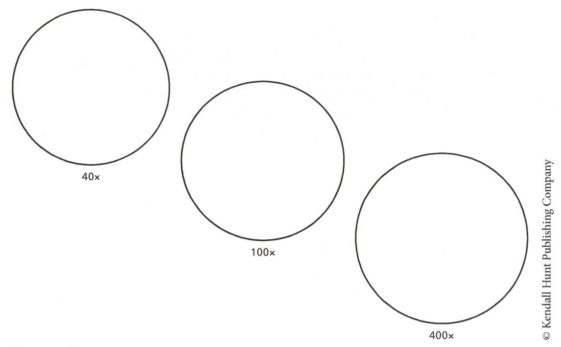

40×

100×

400×

Figure 4.4

Letter e Slide

1. Hold the slide up to the light; the specimen is visible with the naked eye.

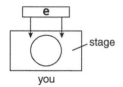

Figure 4.5

2. Orient the slide so that the letter 'e' is right-side up. Place the slide on the stage this way.
3. Using the correct procedure, focus on the specimen at 40× total magnification.
   a. The mirrors in the scope flip and reverse the image before it reaches your eyes.
   b. Compared to the right side up 'e' you saw with the naked eye, the magnified image is _____

   _____

   c. Draw it to scale in Figure 4.6.
4. *When you move the stage in one direction, the image moves in the other direction.*
   a. But you won't really need to think about this as you proceed; your brain will adapt as you use the scope.
5. Focus at 100 ×.
   a. Sketch what you see, to scale, in Figure 4.6.
6. Focus at 400 × total magnification.
   a. Can you see the entire letter? _____
   b. Sketch what you see, to scale, in Figure 4.6.

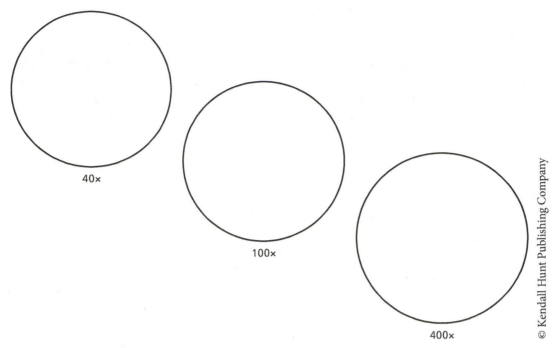

Figure 4.6

Estimating the Size of a Specimen

Since we know the diameter of the field of view (see Table 1), it is easy to *estimate* the size of a specimen. There are two approaches, *which are equivalent*. You should get the same answer using both approaches.

1. What proportion of the diameter of the field of view does the specimen occupy? Multiply that proportion by the diameter of the field of view. See Figure 4.7.
   a. For example, at 40×, if specimen takes up half the diameter of the field of view, it is [(0.5)*(5 mm)] or 2.5 mm in size.

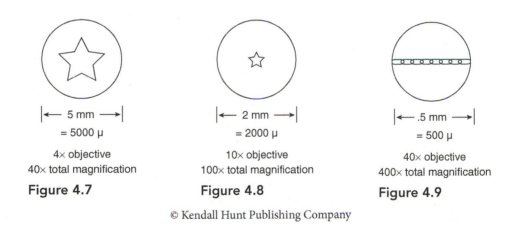

Figure 4.7   **Figure 4.8**   **Figure 4.9**

© Kendall Hunt Publishing Company

2. How many of the specimens could fit across the diameter of the field of view? Divide the diameter of the field of view by that number. (Still referring to Figure 4.7.)
   a. For example, at 40×, if the specimen occupies half of the field of view, then two objects the size of the specimen could fit across the diameter of the field of view. It is [(5 mm)/2] = 2.5 mm in size.

Practice

Look at Figure 4.8 showing a specimen as viewed at 100× total magnification.

1. Estimate its size using approach 1: The specimen viewed at 100× takes up 20 percent of the field of view.
   a. Its estimated size is _____ mm.
   b. Show how you arrived at your answer. _____
   c. Convert this to µ. _____ µ
      (See the size of things.)
2. Estimate its size using approach 2: Five objects the size of the specimen could fit accross the diameter of the field view.
   a. Its estimated size is _____ mm.
   b. Show how you arrived at your answer. _____
   c. Convert this to µ. _____ µ

Look at Figure 4.9 showing a specimen as viewed at 400× total magnification.

3. At 400×, eight objects the size of the specimen could fit across the diameter of the field of view. In other words, the specimen takes up about 12% of the field of view. Use either method to estimate the size of the specimen.
   a. The estimated size is _____ mm.
   b. Show how you arrived at your answer. _____
   c. Convert this to µ. _____ µ

You must use the appropriate magnification to make your estimate. If the specimen takes up most of the field of view, use the next lower magnification. If the specimen is so small that it is difficult to estimate how many would fit, or what proportion of the field of view it occupies, use the next higher magnification. You may not be able to reasonably estimate the size of a specimen. (This is the case for the Eubacteria and Archaea cells which are 1–10 µ.)

CELLS

Prokaryotes

The term *prokaryote* means "before kernel" and refers to cells that do not have a nucleus. All prokaryotic cells are single-celled, and in addition to lacking a nucleus, they also lack much else in the way of internal organization. The term *prokaryote* generally encompasses two kingdoms, the Eubacteria and the Archaea. We will look at examples only of Eubacteria.

Observe the three demonstration microscopes.

1. Prepared slides of bacteria are already set up at 400×.
2. If wells are not already made, use a pipette to carefully remove a plug of agar from the center of each side of the plate.
3. You may need to adjust the fine focus in order to discern the shapes.
4. DO NOT move the coarse focus, the objective lens, or the stage.
5. The cells are small enough (1–10 µ) that you will not be able to accurately estimate the size or make out details (or lack thereof, i.e., little or no internal organization).
6. Complete the table below.

TABLE 4.2 The Three Major Bacterial Shapes			
Shape	Coccus	Bacillus	Spirillum
Written description	spherical	rod-shaped	spiral
Drawing (not to scale)			

Eukaryotes

Eukaryotic organisms may be single-celled or multi-celled. They have, in addition to the nucleus, several other membrane-bounded organelles, which are specialized structures that perform specialized functions. Examples include mitochondria and chloroplasts. You will easily see chloroplasts in the aquatic plant *Elodea*. But mitochondria are small enough that, even when dyed, it takes time and effort to observe them with the light microscope. Eukaryotic cells also have an extensive cytoskeleton made of linear protein elements. You will not be able to directly observe the cytoskeleton today, but will see evidence of it if you observe cytoplasmic streaming—the flowing of cytoplasm that enables the movement of molecules within the cell. Eukaryotes include the kingdoms Fungi, Plants, Animals, and Protists. To look at an example of each of these groups, you will make wet mounts.

HOW TO MAKE A WET MOUNT

1. Place a drop of suspension (if fungus or protist) or a drop of water and then the specimen (if plant or animal) on the slide.
2. Place the cover slip as shown below. Slowly lowering the cover slip will take advantage of the adhesive and cohesive properties of water to minimize air bubbles.
   a. Air bubbles large enough to be seen with the naked eye may impede your ability to view your specimen. Gently tap on the cover slip to remove them, or remove the cover slip and try again.
   b. Small air bubbles, which may appear as very regular shapes with a dark edge, may be annoying, but you should be able to work around them.
   Before you place the slide on the stage, make sure the bottom of the slide is dry.
3. If, when you view your wet mount, it appears to be vibrating, remove the slide and use a small piece of paper towel to wick up some of the excess water.
4. When you are finished with the wet mount, rinse and dry the slide and cover slip.

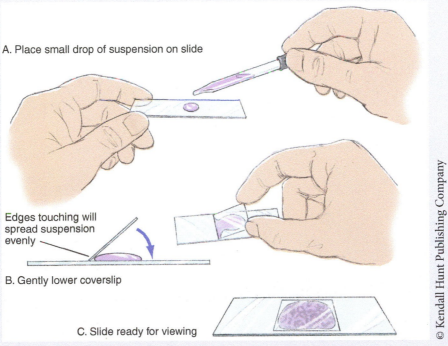

A. Place small drop of suspension on slide

Edges touching will spread suspension evenly

B. Gently lower coverslip

C. Slide ready for viewing

© Kendall Hunt Publishing Company

Figure 4.10 How to make a wet mount.

PLANT, ANIMAL, FUNGUS, PROTIST

For each of the live samples, *use the most appropriate magnification* to:

1. Make a drawing (to scale; draw what you see).
2. Estimate the size of a single cell.
3. Give a written description of the shape of the cells.

Plant

1. Make a wet mount of the smallest developing leaves at the growing tip (apical meristem) of the freshwater aquatic plant *Elodea*.
2. In your drawing, label the cell wall and chloroplasts.

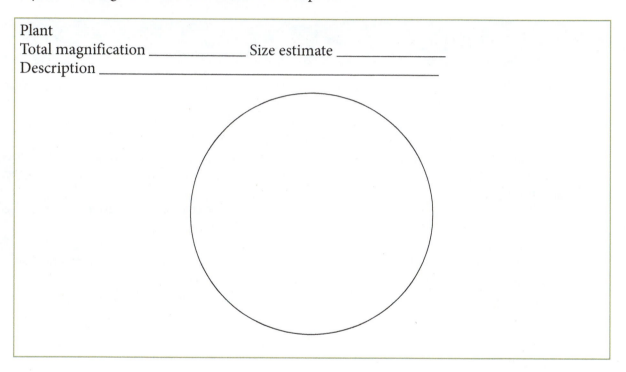

Plant
Total magnification _____ Size estimate _____
Description _____

Animal

1. Make a wet mount of your cheek epidermal cells.
    a. Use a toothpick to gently scrape the inside of your cheek to remove cells that have already sloughed off or are ready to slough off. DO NOT break or tear the skin.
    b. After you have transferred the cells from the toothpick to the slide, dispose of the toothpick in the disposal container provided. DO NOT lay your used toothpick on the bench.
2. Stain the specimens by adding a small drop of methylene blue. This will enable you to more easily see the cells and the nucleus in each cell.
3. In your drawing label the nucleus and cytoplasm.

Animal

Total magnification _____ Size estimate _____

Description _____

Fungus

1. Make a wet-mount of the suspension containing *Saccharomyces cerevisiae*, commonly called baker's yeast. (Many fungi, including mushrooms, are macroscopic.) You can add a drop of water to your slide to thin the suspension.

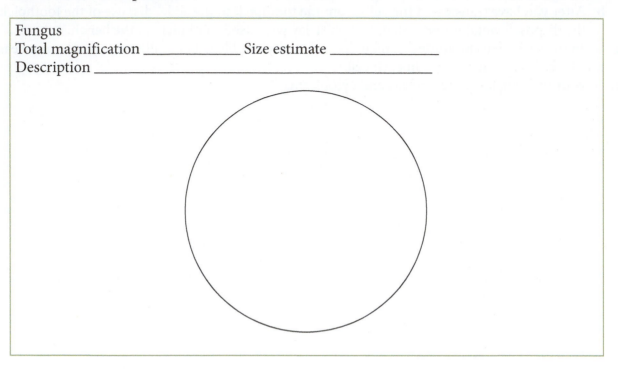

Fungus
Total magnification _____ Size estimate _____
Description _____

Protist

1. View a prepared slide of a protist specimen.
2. Include as much detail as possible in your drawing.

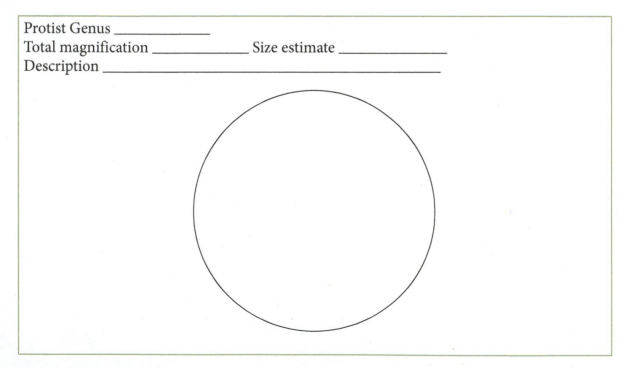

Protist Genus _____
Total magnification _____ Size estimate _____
Description _____

When You Are Done with the Microscope

1. Put slides away as directed.
2. Put the scope away properly.
   a. Make sure there are no slides on the stage.
   b. Position the stage at its midpoint.
   c. Use the revolving nosepiece to engage the 4× objective.
   d. Wrap the electrical cord securely around the body of the scope.
   e. Put on the dust cover.
   f. Place carefully in microscope cabinet with handhold on arm easily accessible for next user.

CELL COMPONENTS

The light microscope enables you to see both prokaryotic and eukaryotic cells, and to see some structures within eukaryotic cells. But there are many structures that cannot be easily visualized. It is important to have a general knowledge of cell structures and their functions. Use Table 3, to answer the questions below.

Which components are common to all cells?

_____

_____

_____

Which kingdoms are eukaryotic?

_____

_____

_____

_____

Which components are found in plants but not animals?

_____

_____

_____

TABLE 4.3 Cell Components	Prokaryote	Eukaryote			
		Protists	Fungi	Plants	Animals
Cell wall	✓	✓ (many)	✓	✓	
Cell or plasma membrane	✓	✓	✓	✓	✓
Ribosomes	✓	✓	✓	✓	✓
DNA	✓	✓	✓	✓	✓
Membrane-bounded organelles		✓			
Nucleus		✓	✓	✓	✓
Endomembrane system (rough smooth ER, vesicles, Golgi complex)		✓	✓	✓	✓
Mitochondria		✓	✓	✓	✓
Chloroplasts		✓ (some)		✓	
Central vacuole				✓	
Cytoskeleton		✓	✓	✓	✓

View the general animal cell model and identify the cell components represented. Research these components to complete the table below.

TABLE 4.4 Animal Cell Model	
Cell Component	**Information such as primary function, basic description, etc.**
Plasma membrane	
Cytoplasm	
Mitochondria	
Ribosomes	
Rough ER	
Smooth ER	
Golgi complex	
Nucleus	
Nucleolus	
Nucleoplasm	
Centrioles	

The Size of Things

$1 \text{ cm} = 10^{-2} \ (1/100) \text{ m}$
$1 \text{ mm} = 10^{-3} \ (1/1000) \text{ m}$
$1 \ \mu\text{m} = 10^{-6} \ (1/1,000,000) \text{ m}$
$1 \text{ nm} = 10^{-9} \ (1/1,000,000,000) \text{ m}$
100 cm per meter
10 mm per cm
1000 μm per mm

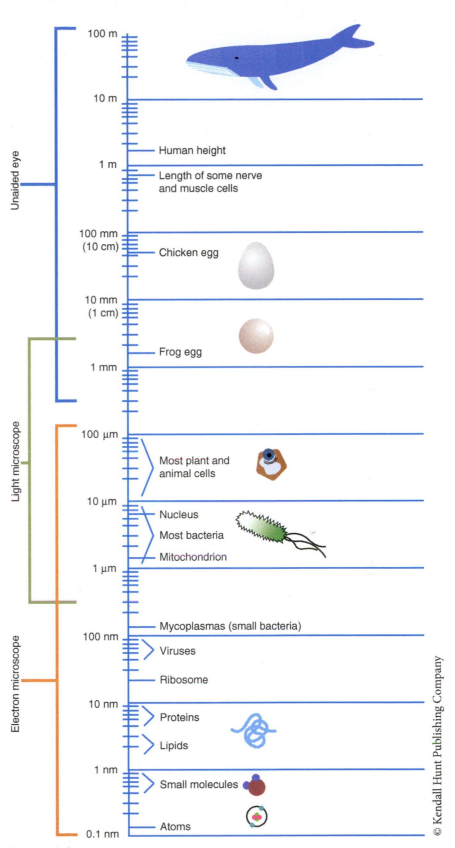

Figure 4.9
© Kendall Hunt Publishing Company

LAB 5

Transport

The plasma membrane is selective about what passes through it because of its molecular composition. It allows nutrients and oxygen to enter the cell but keeps out many substances. By the same token, valuable cell proteins and other substances are kept within the cell, while wastes pass to the exterior. This property is known as differential, or selective, permeability.

All molecules possess kinetic energy and are in constant motion. Since kinetic energy is directly related to both mass and velocity, **smaller molecules tend to move faster**. As molecules move about randomly at high speeds, they collide and ricochet off one another, changing direction with each collision. Although individual molecules cannot be seen, the random motion of small particles suspended in water can be observed. This is called Brownian motion.

CELL MEMBRANE

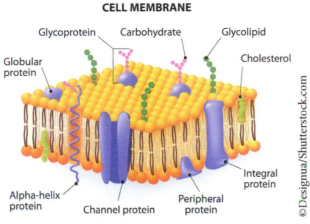

Figure 5.1 Cell Membrane Structure

DIFFUSION

Diffusion is the net movement of molecules from a higher concentration to a lower concentration. Diffusion will occur until a dynamic equilibrium is achieved and the molecules are equally distributed. The driving force of diffusion is the kinetic energy of the molecules themselves. The speed of diffusion is dependent on such factors as temperature, the size of the molecules, the type of medium, and the concentration gradient.

The diffusion of particles into and out of cells is modified by the plasma membrane. Small, nonpolar molecules can pass through the phospholipid bilayer with relative ease. Small, polar molecules usually need to cross at a protein ion channel in the membrane. **Simple diffusion** is the diffusion of solutes across a differentially permeable membrane.

Diffusion of Dye through Agar Gel

The relationship between molecular weight and the rate of diffusion can be examined easily by observing the diffusion of the molecules of two different types of dye through an agar gel. The dyes used in this experiment are methylene blue and potassium permanganate. Methylene blue has a molecular weight of 320 atomic mass units (amu) and is deep blue. Potassium permanganate has a molecular weight of 158 amu and is purple. Although the agar gel appears quite solid, it is approximately 98.5% water and allows free movement of the dye molecules through it.

1. Locate methylene blue and potassium permanganate dyes. Double check the LABELS!!!
2. If wells are not already made, use a pipette to carefully remove a plug of agar from center of each side of the plate.
3. Carefully place a small drop of methylene blue dye in one well in the agar.
4. Fill the remaining well with a small drop of potassium permanganate dye.
5. Note and record your starting time.
6. Immediately measure the diameter of each circle of dye when you return to your table.
7. Record this first measurement as zero time, and place the cover on the Petri dish.
8. Continue to measure the diameter, in a consistent manner, of each circle every 15 minutes over the next hour.

TABLE 5.1 Investigation of Molecular Size on Rate of Diffusion		
Time	Diameter of the Methylene Blue (millimeters)	Diameter of the Potassium Permanganate (millimeters)
Zero (initial measurement)		
15 minutes		
30 minutes		
45 minutes		
60 minutes		

QUESTIONS

1. Which dye diffused more rapidly? _____

2. What is the relationship between the molecular weight and the rate of molecular movement? _____
_____
_____

3. Why did the dyes move? _____
_____
_____
_____

4. Compute the rate of diffusion and record.

Diffusion rate = Millimeters of movement / Time in minutes.

A. Potassium Permanganate _____ mm/min

B. Methylene Blue _____ mm/min

5. How would these rates be affected if the gels were made harder by adding more agar?

_____

OSMOSIS

Osmosis is a specific type of diffusion; it is the diffusion of water molecules across a semi-permeable membrane. Because it is a type of diffusion, the net movement will still occur from high to low concentration and will end at dynamic equilibrium. The fluid surrounding cells contain small amounts of dissolved substances (solutes) that are equal to, less than, or greater than those found within the cell.

The relationship between the concentrations of solutes on either side of the membrane is referred to as tonicity and each side can be described by one of three terms: isotonic, hypertonic, or hypotonic.

In isotonic relationships, there is no net movement of water between cell and environment. The concentration of solutes is the same on either side of the membrane.

The side of the membrane with a higher concentration of solute is called hypertonic. A cell in a hypertonic solution will lose water.

The side of the membrane with a lower concentration of solute is said to be hypotonic. A cell in a hypotonic solution will gain water.

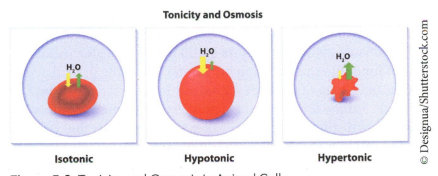

Figure 5.2 Tonicity and Osmosis in Animal Cells

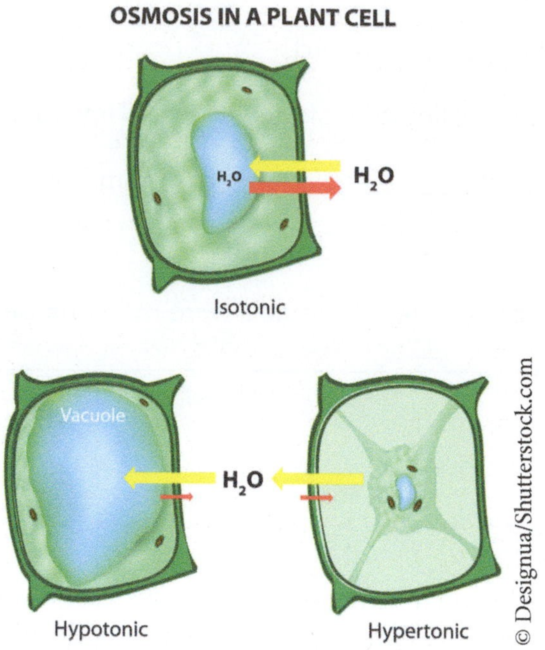

OSMOSIS IN A PLANT CELL

Figure 5.3 Tonicity and Osmosis in Plant Cells

Observing Diffusion and Osmosis across a Membrane

The following experiment provides information on the movement of water and solutes through differentially permeable membranes.

1. Obtain five dialysis sacs, ten clamps, a funnel, a graduated cylinder, masking tape, and five small beakers.
2. Label the small beakers 1, 2, 3, 4 and 5 using the masking tape.
3. Prepare the dialysis sacs one at a time.

For sac1, obtain a piece of dialysis tubing, soaking in distilled water, from the fingerbowl. **Fold one end of the dialysis tubing twice and apply a clamp**. Double folding helps decrease the possibility of leaks. Open the unclamped end of the dialysis tubing; insert the funnel into the sac to keep the sac open. Using the large graduated cylinder, measure 20 ml of the 40% glucose solution and then pour the measured amount of fluid into the sac using the funnel. Carefully press the air out of the sac and **double fold the opening** and clamp the dialysis sac. Gently blot the sac dry using a paper towel. Weigh the sac on the scale and record the weight on the data sheet. Place the sac onto a labeled paper towel. Put approximately 150 mL of distilled water into beaker 1. Assign a member of your team to rinse the funnel and the graduated cylinder at the sink with water and dry.

4. Repeat the above process with each of the remaining pieces of dialysis tubing substituting the appropriate sac fluids and beaker fluids as follows:

   Sac 2: Fill sac 2 with 20 ml of 40% glucose. Put approximately 150 mL of 40% glucose solution into beaker 2.
   Sac 3: Fill sac 3 with 20 ml of 5% NaCl solution. Put approximately 150 mL of distilled water into beaker 3.

Sac 4: Fill sac 4 with 20 ml of 40% sugar solution + Congo Red dye. Please note that this has already been mixed for you. Simply measure 20 mL of the red solution and add it to sac 4. Put approximately 150 mL of distilled water into beaker 4.

Sac 5: Fill sac 5 with 20 ml of distilled water. Put approximately 150 mL of 40% glucose solution into beaker 5.

5. Once all sacs are ready, place them into their respective beakers and record the start time.

Allow the sacs to remain undisturbed in the beakers for 45 minutes.

While you are waiting for the sacs, it is time to make a few predictions about the osmosis part of the experiment. Decide the type of environment (hypertonic, hypotonic, or isotonic) in which each sac was placed. Based on the environment, make a prediction about what will happen to the weight of the sac.

TABLE 5.2 Predictions about Tonicity			
Sac	Contents of Sac and Environment/ Beaker contents	Environment (hyper-, hypo-, or isotonic)	Predicted Weight Change (increase, decrease, stay the same)
1	40% glucose solution IN distilled water		
2	40% glucose solution IN 40% glucose solution		
3	5% NaCl solution IN distilled water		
4	40% sugar solution with Congo Red dye IN distilled water		
5	Distilled water IN 40% glucose solution		

Osmosis might not be the only process at work in this experiment. It is time to consider other factors that may influence the results.

In sac 1, would the glucose move into or out of the sac in a net fashion if it can pass the selectively permeable membrane? _____

What would this process be called?_____

What effect would this have on the weight of the bag?_____

In sac 2, would the glucose move into or out of the sac in a net fashion if it can pass the selectively permeable membrane? _____

Why or why not? _____

In sac 3, would the salt move into or out of the sac in a net fashion if it can pass the selectively permeable membrane? _____

What would this process be called?_____

What effect would this have on the weight of the bag?_____

In sac 4, would the sugar move into or out of the sac in a net fashion if it can pass the selectively permeable membrane? _____

What would this process be called?_____

What effect would this have on the weight of the bag?_____

In sac 4, would the red dye move into or out of the sac in a net fashion if it can pass the selectively permeable membrane? _____

What would this process be called?_____

What effect would this have on the weight of the bag?_____

In sac 5, would the sugar move into or out of the sac in a net fashion if it can pass the selectively permeable membrane? _____

What would this process be called?_____

What effect would this have on the weight of the bag?_____

6. After approximately 30 minutes, prepare a water bath. Obtain a large beaker and add approximately 200 mL of distilled water. Place the beaker on the cold hotplate, and turn the hotplate on to high/300°C. Bring the distilled water to a boil.

7. After the 45 minutes have passed (or on the request of the instructor) remove the sacs from the beakers. Gently blot the sacs dry and weigh the sacs on the same scale used previously. Record the weights in Table 3. Place the sacs back onto their labeled paper towels (1, 2, 3, 4, and 5). Place the appropriate sac on the labeled paper towel. See the appropriate section for each sac for further instructions.

TABLE 5.3 Investigation of Diffusion and Osmosis across a Membrane						
Beaker Filled With:	Contents of Sac	Initial Weight	Final Weight	Weight Change	Tests: Beaker	Test: Sac Fluid
Beaker 1 150 ml distilled water	Sac 1 20 ml 40% glucose solution				Benedict's Test:	Benedict's Test:
Beaker 2 150 ml 40% glucose solution	Sac 2 20 ml 40% glucose solution					
Beaker 3 150 ml distilled water	Sac 3 20 ml 5% NaCl solution				AgNO$_3$ Test:	
Beaker 4 150 ml distilled water	Sac 4 20 ml 40% sugar solution with Congo Red dye				Benedict's Test:	
Beaker 5 150 ml 40% glucose solution	Sac 5 20 ml distilled water					Benedict's Test

Sac 1

8. As indicated in the above chart, you are going to perform the Benedict's Test for reducing sugars on the solution in the beaker and the solution in the sac. Obtain two large test tubes. Label one of the tubes 'sac' and the other tube 'beaker'. Place 2 ml of Benedict's Solution in each test tube. Carefully

remove one of the clamps from sac 1. Using a graduated pipette, obtain 4 ml of fluid from inside of sac 1 and place this liquid into the test tube labeled 'sac'. Using a clean pipette, likewise pipette 4 ml of fluid from the beaker labeled 1. Place the 4 ml of beaker fluid into the test tube labeled 'beaker'. Place both test tubes into the boiling water bath. Boil for three minutes. If a green, yellow, orange, or rusty precipitate forms, the test is positive for the presence of glucose. If the solution remains blue (the original color) the test is negative for glucose. Indicate the results in Table 3.

QUESTIONS

1. Was there a change in weight? If there was a change of weight, explain why the weight changed.

_____

_____

_____

_____

2. Was glucose still present in the sac? _____

3. Was glucose present in the beaker? _____

4. Why was glucose present in the beaker?

_____

_____

Sac 2

9. Sac 2 does not have any further chemical tests. Assist your lab group members with their tests.

QUESTIONS

1. Was there a significant change in weight? _____

2. With 40% glucose in the sac and the beaker, would you expect to see any net movement of water or glucose molecules? _____

3. Explain your answer to number 2.

_____

_____

Sac 3

10. Obtain one large test tube. Label the test tube beaker 3. Using your pipette, measure 3 ml of fluid from beaker 3 and pour this into the test tube. Add one drop of $AgNO_3$ (silver nitrate) to the test tube labeled beaker 3. The appearance of a white precipitate or cloudiness indicates the presence of AgCl. AgCl is formed by the reaction of $AgNO_3 + NaCl \rightarrow AgCl + NaNO_3$. Record the results in Table 3.

QUESTIONS

1. Did the weight of sac 3 change? If so, explain why. _____
_____
_____
_____
_____

2. Is there a precipitate (i.e. does the solution appear cloudy)? _____

3. If there is a precipitate, explain why. _____
_____
_____
_____

Sac 4

11. Obtain a large test tube and label the test tube beaker 4. Pipette 4 ml of beaker fluid and place this fluid into the test tube. Add 2 ml of Benedict's solution to the test tube and place this tube into a boiling water bath. Boil the solution for three minutes. If a green, yellow, orange, or rusty precipitate forms, the test is positive for the presence of a reducing sugar. If the solution remains blue (the original color) the test is negative for a reducing sugar. Record your results in Table 3.

Congo Red, a biological stain/indicator, has a molecular weight 696.67 amu.

QUESTIONS

1. Did the Congo Red dye diffuse causing the water to turn pink? _____

2. If the Congo Red dye did not diffuse explain why.

_____

_____

3. Has there been a weight change? If so, explain why the weight changed.

_____

_____

_____

4. Did sugar diffuse from sac 4 into the beaker fluid? _____

Sac 5

12. Obtain a large test tube and label the test tube sac 5. Pipette 4 ml of sac fluid and place this fluid into the test tube. Add 2 ml of Benedict's solution to the test tube and place this tube into a boiling water bath. Boil the solution for three minutes. If a green, yellow, orange or rusty precipitate forms, the test is positive for the presence of a reducing sugar. If the solution remains blue (the original color) the test is negative for a reducing sugar. Record your results in Table 3.

QUESTIONS

1. Did the weight of sac 5 change? If so, explain why.

_____

_____

_____

2. Was glucose present in the sac? If so, explain why. _____

_____

Sacs 1 to 5

1. In which of the test situations did net simple diffusion occur?

   _____

   _____

2. In which of the test situations did net osmosis occur?

   _____

   _____

3. With what cell structure can the dialysis sac be compared?

   _____

4. Unlike a dialysis sac, what might happen to your cells if they take on too much water by osmosis?

   _____

LAB 6

Photosynthesis

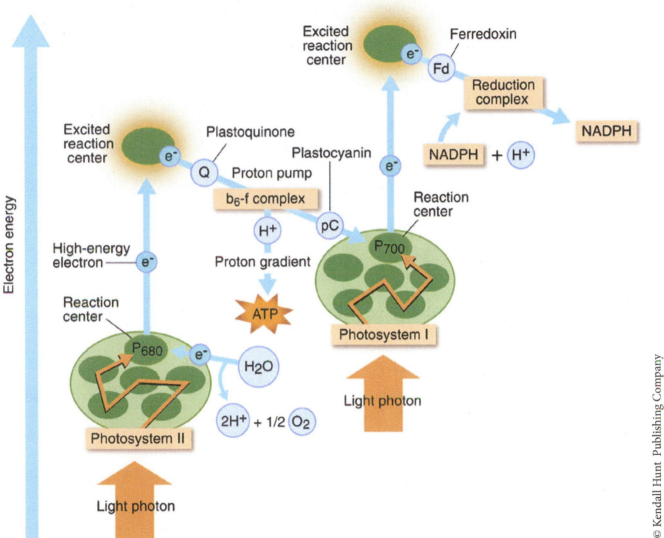

Figure 6.1 Noncyclic pathway of the light dependent reactions of photosynthesis.

Adapted from *Investigating Biology: The Unity of Life Lab Manual* by Paul Florence and Anissa Florence. Copyright © 2013 by Paul Florence and Anissa Florence. Reprinted by permission of Kendall Hunt Publishing Company.

© Kendall Hunt Publishing Company

Paper chromatography is a common laboratory method used to separate compounds from a mixture. Paper chromatography can be used to separate out such compounds as amino acids, pigments, and dyes. We will use this procedure to separate out photosynthetic pigments from spinach leaves. The pigments we will study are **chlorophyll a**, **chlorophyll b**, **beta-carotene**, and two different **xanthophylls**.

One of the main characteristics of each plant pigment is its solubility in the solvent we are using, acetone. Acetone is a nonpolar solvent, so pigments that are more nonpolar have a higher solubility than those that are more polar. The pigments will separate out of the leaf extract solution based on their solubility in the solvent, which will create different colored pigment bands on the chromatography paper.

After the pigments are separated out, the R_f (relative front) value is calculated. R_f values can be useful in determining unknown pigments. The R_f value is calculated with the following formula:

$$R_f = \frac{\text{Distance the pigment band migrated}}{\text{Distance the solvent front migrated}}$$

Extraction of Spinach Leaf Pigments

1. Obtain two spinach leaves and remove the veins (you want to use only the fleshy parts of the leaves).
2. Place the leaves into a mortar and add 20 ml of cold acetone.
3. Grind the leaves with the pestle until the leaves are pulverized and the acetone has turned a dark green color.
4. Pour the leaf extract into a large test tube and allow the pulp to settle (this will take a couple of minutes).
5. Fill a glass beaker with lid with enough chromatography solvent (9 parts petroleum ether: 1 part acetone) to cover the bottom of the container. (Fill just enough to cover the bottom. If you use too much solvent in the beaker it will just wash the pigments out of your paper.)
6. Place lid on the container. This will saturate the atmosphere of the beaker with the solvent.
7. While you are waiting for the leaf pulp to settle, use soap to wash the mortar and pestle. If there is still a lot of green color remaining, you can use some of the left over acetone to wash the mortar and pestle.

Separating Pigments from the Extract

1. Transfer the top portion of the extract (it will be a darker green color) to a small test tube. You will use extract from this small test tube for your experiment.
2. Obtain a piece of chromatography paper from your instructor. Minimize how much contact your fingers have with the paper, as oils from your skin can affect the experiment.
3. Make a light pencil mark on your paper approximately 1 cm from the bottom. This will designate the point of origin for your pigments.
4. Using a capillary tube streak your pigment extract across the pencil line 7 times, allowing the streak to dry in between each application. Try to keep the pigment line as narrow as possible; if the extract spreads out to form a wide band the experiment will not work as well.
5. Allow the pigment extract band to dry completely before putting the paper into the container.
6. Make sure the volume of solvent in your container is not deeper than the height of your pigment line. If the solvent volume is acceptable, uncover the container and place the paper into the container. Re-cover the container with the lid.

7. The solvent will move up the paper via capillary action. When the solvent front gets approximately 1 cm from the top of the paper remove the paper from the container (this usually takes around 20–45 minutes).
8. Immediately mark, on the edge of the paper, how far the solvent front moved.
9. You should immediately measure (in centimeters) how far each pigment band traveled (from origin to bottom of pigment band).
10. Once you have marked your solvent front and measured pigment bands, make a sketch of your paper noting the position and color of each pigment band.
11. This "developed" paper is called a chromatogram.

See Figure 6.2 below for a summary of paper chromatography and an illustration of what your chromatogram should look like.

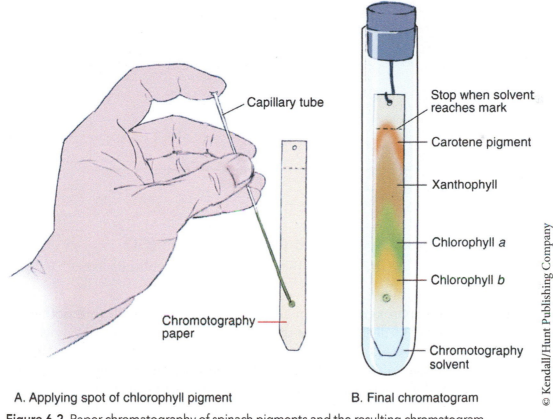

A. Applying spot of chlorophyll pigment

B. Final chromatogram

Figure 6.2 Paper chromatography of spinach pigments and the resulting chromatogram.

Draw and label your chromatogram in the box provided.

The Relationship of Photosynthesis to Carbon Dioxide (CO_2)

This procedure will demonstrate the fact that the process of photosynthesis uses CO_2 from the surrounding environment.

1. Fill a small beaker with approximately 60 ml of a dilute solution of BTB (bromothymol blue), a pH indicator solution. BTB will appear blue-green when the solution has a pH above 6.6 and pale yellow when the pH is below 6.0.
2. With a straw *gently* blow air into the test tube until the solution changes to yellow. The CO_2 you are breathing into the solution reacts with water to form carbonic acid which causes the color change.
3. Add equal volumes of yellow BTB into three large test tubes. Leave room for the addition of a plant.
4. Place a small *Elodea* branch in two of the test tubes.
5. Cap all tubes.
6. Cover one of the tubes containing *Elodea* with aluminum foil so that all light is excluded.
7. Put all three tubes in bright light for 60 minutes.

8. Record any color changes in the table below.

Test Tube	Initial Color	End Color
Test Tube 1 (no *Elodea*)		
Test Tube 2 (with *Elodea*)		
Test tube 3 (with Elodea, no light)		

Review Questions

1. Complete the following table using your data.

Pigment R_f	Pigment Name	Pigment Color

2. Which pigment in spinach leaves has the highest solubility in the solvent?

____

3. Which pigment(s) in spinach leaves has the lowest solubility in the solvent?

____

4. Which pigment is the most polar?

____

5. Why did the solution in test tube 2 change color?

____

____

6. Why do plants need CO_2?

____

7. If bubbles are present on the *Elodea* leaves in test tube 2 what gas are these bubbles made of?

____

8. How is oxygen formed during the light reactions?

____

____

9. What color is the fluid in Test Tube 3 (Elodea, no light)?

_____

10. Did carbon fixation occur in Test tube 3?

_____

PROCESS OF PHOTOSYNTHESIS

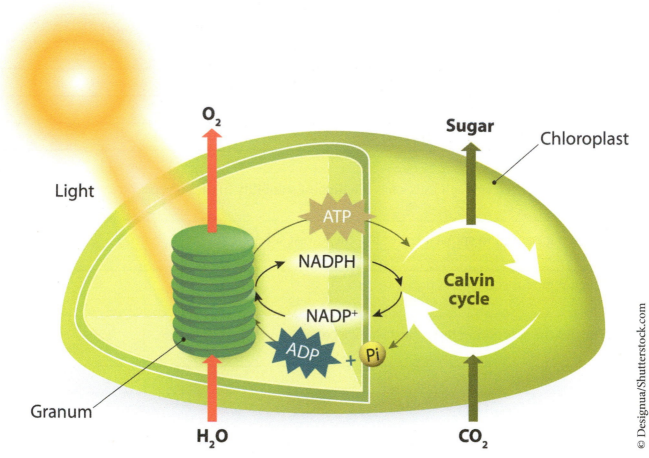

Figure 6.3 Overview of Photosynthesis

Name (s): _____   Class: _____   Score _____

LAB 7

Cellular Respiration

The transfer of energy in **aerobic** cellular respiration can be represented in the following equation:

$$C_6H_{12}O_6 + 6O_2 \rightarrow 6CO_2 + 6H_2O + 36\text{-}38 \text{ ATP}$$

In order to maintain the organization of life, organisms must have a constant input of energy. This energy is usually obtained in the form of glucose ($C_6H_{12}O_6$) and then transferred into a more usable form of energy called ATP. As explained by the second law of thermodynamics, some of this energy is also lost to the environment as heat. Aerobic cellular respiration begins with glycolysis in the cytoplasm of a eukaryotic cell, with the final two stages taking place in the mitochondria. The Kreb's cycle takes place in the mitochondrial matrix, and the electron transfer system takes place in the cristae.

ANATOMY OF A CELL

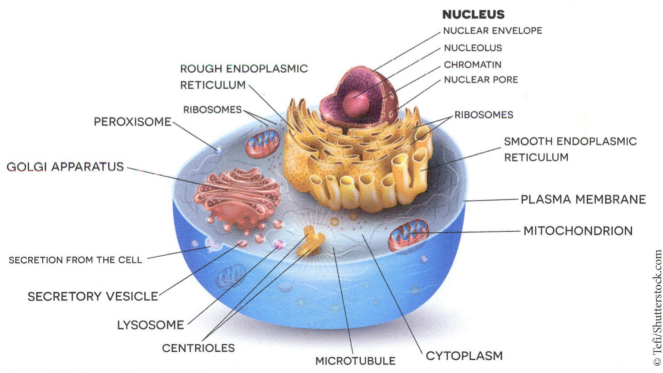

© Tefi/Shutterstock.com

Figure 7.1 General Animal Cell Structure

Adapted from *Investigating Biology: The Unity of Life Lab Manual* by Paul Florence and Anissa Florence. Copyright © 2013 by Paul Florence and Anissa Florence. Reprinted by permission.

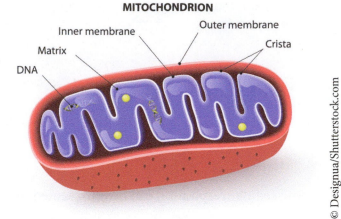

Figure 7.2 Mitochondrial Structure

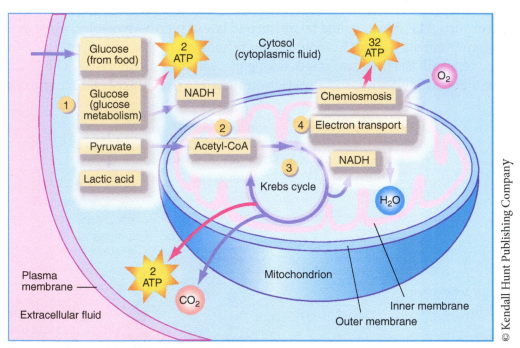

Figure 7.3 Aerobic Respiration Pathway

EXERCISE AND OXYGEN

In this activity, you will explore the relationship between exercise and the rate of respiration in humans. Look over the aerobic respiration equation and think about your own life experiences to develop a hypothesis about this relationship. Record this hypothesis below.

1. Hypothesis: _____

_____

_____

Work with your lab partners to develop the protocols for the experiment you will conduct to investigate the relationship between exercise and respiration rate. Before you begin the experiment, discuss your experimental design with your instructor. Your instructor will be particularly interested in how you will obtain the data for the control group, what exercise will be performed, for how long the exercise will be performed, who will exercise, who will measure the respiration rate, who will time the exercise and respiration measurements, and how you decided who will exercise. For data consistency, respiration rates should be measured as the number of inhalations in the one minute immediately following exercise.

2. Now, rewrite the hypothesis as a prediction statement specific to your experimental design:

Prediction Statement:

If _____

_____ then _____

_____

Conduct your experiment and write your results on the board. Record the data for the entire class in Table 1 of your handout.

TABLE 7.1 Class Data on the Effect of Exercise on the Respiration Rate of Humans		
Respiration (inhalations/minute)		
	Before	**After**
1.		
2.		
3.		
4.		
5.		
6.		
7.		
8.		
9.		
10.		
11.		
12.		
13.		
14.		
15.		
16.		
17.		
18.		
19.		
20.		
Total		
Average		

3. Based on the class results, do you accept or reject your hypothesis? Be sure to write the answer as a full sentence conclusion. _____

_____

4. What were some potential errors that could have affected the class data?

_____

_____

_____

ANAEROBIC RESPIRATION

Not all organisms use aerobic respiration, and even those that do can switch to anaerobic means. In the aerobic respiration equation, you can see that glucose is completely broken down into carbon dioxide and water. In fermentation pathways, glucose will not be completely broken down, leaving a lot of energy in the products. Many animal cells can resort to lactic acid/lactate fermentation when their cells run low on oxygen. Later the animal can convert the lactate into a usable form of energy (pyruvate) to extract the remaining energy. Some bacteria and fungi can also use lactic acid fermentation. Fermentation pathways take place in the cytoplasm of the cell.

Some bacteria and fungi (primarily yeasts), undergo alcohol/ethanol fermentation. This type of fermentation is commercially valuable to humans in the baking and brewing industries. The CO_2 released from yeast during the baking process causes bread and other baked goods to rise. The alcohol is burned off during the baking process.

Yeasts are also used to make beer, wine, and other spirits. The ethanol is preserved as an important ingredient. It should be noted that ethanol is a by-product of the fermentation pathway, and it is toxic to the yeast that produce it. This limits the amount of alcohol to about 12% in a naturally fermented product.

Alcohol/Ethanol Fermentation Equation: $C_6H_{12}O_6 \rightarrow 2CO_2 + 2C_2H_5OH + 2 \text{ ATP}$

© aboikis/Shutterstock.com

Figure 7.4 Ethanol Fermentation is used in the Production of Wine and Bread.

Alcohol/Ethanol Fermentation by Yeast

1. Measure out 3.5 g of yeast; add to a large beaker.
2. Measure out sugar if required for your flask (see Table 2); add to the beaker.
3. Measure 100 mL of 43°C water in a graduated cylinder & pour into the beaker; stir for 1 minute.
4. Pour beaker contents into an Erlenmeyer flask. Immediately place a balloon over the opening. Use tape to secure the balloon to the flask (see demo if available).
5. Place your flask in the proper location as determined by your group number.
6. Measure the circumference of your balloon in the "center" and record this measurement in the data table. Note starting time.
7. Measure every 15 minutes and record your results in Table 4.

TABLE 7.2 Overview of Experimental Set-Up for First Experimental Group			
Group Number/ Chemistry Box Number	Amount of Yeast	Amount of Sucrose	Location
1	3.5 g	0 g	Table
2	3.5 g	10 g	Table
3	3.5 g	40 g	Table
4	3.5 g	0 g	Water bath
5	3.5 g	10 g	Water bath
6	3.5 g	40 g	Water bath

Replication is important in science. Your group will now set up a flask that is identical to your first flask with the exception of location/temperature. This second flask will allow you to make a direct comparison and serve as a replicate for the class results.

1. Measure out 3.5 g of yeast; add to a large beaker.
2. Measure out sugar if required for your group (see Table 3); add to the beaker.
3. Measure 100 mL of 43°C water in a graduated cylinder & pour into the beaker; stir for 1 minute.
4. Pour beaker contents into an Erlenmeyer flask. Immediately place a balloon over the opening. Use tape to secure the balloon to the flask (see demo if available).
5. Place your flask in the proper location as determined by your group number.
6. Measure the circumference of your balloon in the "center" and record this measurement in the data table. Note starting time.
7. Measure every 15 minutes and record your results in Table 4.

TABLE 7.3 Overview of Experimental Set-Up for Second Experimental Group for Replication			
Group Number/Chemistry Box Number	Amount of Yeast	Amount of Sucrose	Location
1	3.5 g	0 g	Table
2	3.5 g	10 g	Table
3	3.5 g	40 g	Table
4	3.5 g	0 g	Water bath
5	3.5 g	10 g	Water bath
6	3.5 g	40 g	Water bath

Flask	Symbol for Graph	Initial size	15 minutes	30 minutes	45 minutes	60 minutes	75 minutes
1a							
1b							
Average of 1a and 1b	△						
2a							
2b							
Average of 2a and 2b	○						
3a							
3b							
Average of 3a and 3b	□						
4a							
4b							
Average of 4a and 4b	▲						
5a							
5b							
Average of 5a and 5b	●						
6a							
6b							
Average of 6a and 6b	■						

1. Now graph your data. Remember to title your graphs in an informative manner, label your axes properly, provide units of measurement where applicable, and provide a legend. The independent variable is usually placed on the x-axis (horizontal), while the dependent variable is usually placed on the y-axis (vertical).
   a. Make a line graph of all of your data points. Add in lines of best fit.
   b. Make two different bar graphs of your choosing to represent two different experimental treatments represented in the experiment. These must show different interpretations of the data (via means, etc.) than the line graph.

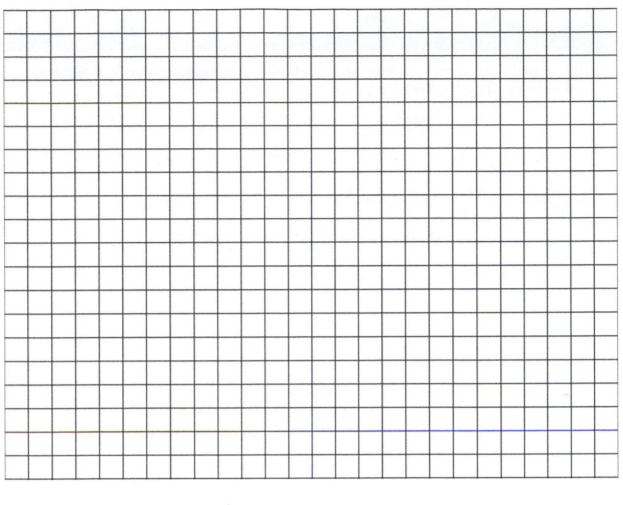

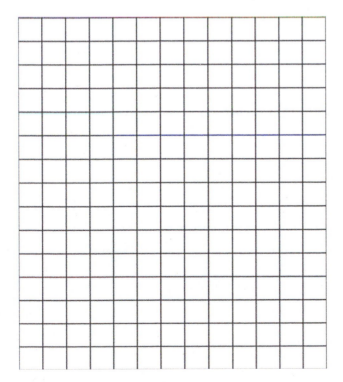

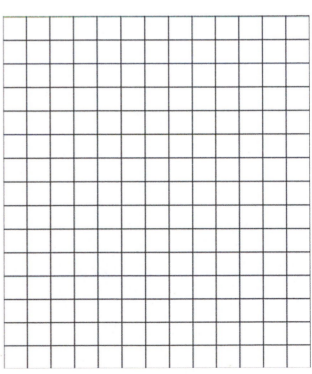

QUESTIONS

1. Which of the set-ups had balloons that reached the largest size on average (list what was placed in the flask and the what temperature it was kept at)? _____
   _____

2. Which of the set-ups had balloons that grew the least on average (list what was placed in the flask and the temperature it was kept at)? _____
   _____

3. What chemical compound caused the balloons to grow in size? _____

4. What were the independent variables? _____
   _____

5. What were the controlled variables in the experiment? _____
   _____

6. Which group(s) served as the control(s)? _____

7. What are the products of alcohol/ethanol fermentation? _____
   _____

8. How could we improve this experiment (be more confident in our results, etc.)?
   _____
   _____

9. Yeast is used to make bread. Using the information you have gained during this lab, explain what causes bread to rise, and explain how you could make the lightest, airiest bread possible. _____
   _____
   _____
   _____
   _____
   _____

LAB 8

DNA Extraction and Protein Synthesis

In 1953, James Watson and Francis Crick deduced the structure of DNA using their own data and those of Rosalind Franklin. DNA is a **double helix**, which looks like a twisted ladder. The outer portions are called sugar-phosphate backbones; they are made of the sugar deoxyribose and the phosphate group of nucleotides. The "rungs" are made of the organic bases of the nucleotides: adenine, thymine, guanine, and cytosine. The adenine (A) of one strand of DNA will base pair in a complementary fashion with thymine (T) on the other strand of DNA through hydrogen bonds to make a "rung." The guanine (G) of one strand of DNA will base pair in a complementary fashion with cytosine (C) on the other strand of DNA through hydrogen bonds to make another "rung." These rules apply to all organisms. Organisms differ from each other in their **DNA nucleotide sequence** only.

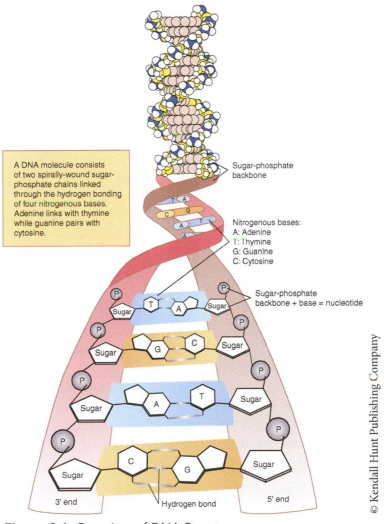

A DNA molecule consists of two spirally-wound sugar-phosphate chains linked through the hydrogen bonding of four nitrogenous bases. Adenine links with thymine while guanine pairs with cytosine.

Sugar-phosphate backbone

Nitrogenous bases:
A: Adenine
T: Thymine
G: Guanine
C: Cytosine

Sugar-phosphate backbone + base = nucleotide

3' end

5' end

Hydrogen bond

© Kendall Hunt Publishing Company

Figure 8.1 Overview of DNA Structure.

EXTRACTING DNA

DNA can be isolated from many organisms using a similar procedure to what you will perform today. As a group, you will prepare a liquid filtrate which contains strawberry DNA in solution. There will be enough filtrate for each person in the group to put some of it into a clean test tube and, following the procedure, extract the DNA as a whitish-looking cloud in a test tube.

1. To a 200 mL beaker, add 90 mL of water. Then, add 10 mL of dishwashing detergent and stir well with a stir rod. This is the DNA extraction liquid, and its components function in important ways:
   - Water in the detergent solution rehydrates cellular material including DNA.
   - Liquid detergent breaks open nuclear membranes to release the DNA.
   - Salt in the solution reacts with the negative phosphate ends of the DNA causing the nucleic acid molecules to stick together; this will enable the DNA to precipitate out when the alcohol is added later.
2. Place two strawberries, without leaves, into a zipper-lock plastic bag.
3. Gently and completely smash the strawberries for about two minutes, working the strawberries between your fingers. This helps start to open the cells and release their DNA.
4. Pour the DNA extraction liquid into the bag with the mashed strawberries. Remove as much air as possible and re-seal the bag.
5. Gently work the strawberries between your fingers for another minute, making sure that all of the smashed strawberries have suspended in the DNA extraction liquid.
6. Place the funnel in the ring stand and drape the cheesecloth over it, using your finger to push the cheesecloth down into the funnel a bit. Place a fresh beaker under the funnel. Pour the strawberry suspension through the cheesecloth, capturing the filtrate in the beaker and holding back the larger, insoluble pieces. You only need 3 mL for each student to advance to the following steps.
7. Using a pipette, each student can measure 3 mL of strawberry filtrate to a test tube. Add 1 mL of meat tenderizer solution (contains proteases) to each test tube and gently mix by swirling. After mixing, allow the solution to rest for 1 minute to allow the protease enzyme time to react with the filtrate. The protein digesting enzymes in the meat tenderizer help clean extraneous protein away from the DNA.
8. Tilt your test tube to a 45-degree angle and **slowly** add 2 mL of ice cold alcohol by trickling it down the inside of the tube to create a layer on top of the cellular suspension.
9. DNA will precipitate out of the solution in a minute or two, as you gently rotate the tube, and become visible as a white stringy mass.

SHOW YOUR INSTRUCTOR YOUR DNA TUBE BEFORE YOU MOVE ON TO THE NEXT ACTIVITY

All of the waste generated is nontoxic and can go down the sink, if liquid, or in the trash, if solid.

PROTEIN SYNTHESIS

This laboratory activity covers the process of protein synthesis. Protein synthesis consists of transcription and translation. Transcription is the activity that produces a copy/version of DNA known as mRNA. Translation is the process of producing the amino acid sequence (polypeptide) from the information in mRNA. Our goal today is to work through the process of translation.

RNA is made of the same subunits (nucleotides) as DNA, but RNA contains the sugar ribose instead of deoxyribose. While, RNA contains adenine, guanine, and cytosine like DNA, it replaces thymine with a base called uracil. Also, RNA is single stranded, and it is not twisted into a helix. Messenger RNA takes the coded message from DNA to the cytoplasm. Messenger RNA complementary base pairs with the DNA sense strand. Transfer RNA brings individual amino acids to the ribosome where they can be connected with peptide bonds to form polypeptides and proteins. Another type of RNA, ribosomal RNA makes up a portion of the ribosomes where proteins are produced. mRNA codes for the amino acids that will make up the primary structure of a protein. It uses 3-letter codons to signal for a particular amino acid; the chart below is used to see which amino acids are encoded. The chart can be used ONLY for mRNA. tRNA consists of anti-codons; 3-letter codes that are the complements of the mRNA codons. The anti-codons will determine which amino acids to pick up for the molecule to bring to the ribosome.

First Base	Second Base: U	C	A	G	Third Base
U	UUU phenylalanine	UCU serine	UAU tyrosine	UGU cysteine	U
	UUC phenylalanine	UCC serine	UAC tyrosine	UGC cysteine	C
	UUA leucine	UCA serine	UAA stop	UGA stop	A
	UUG leucine	UCG serine	UAG stop	UGG tryptophan	G
C	CUU leucine	CCU proline	CAU histidine	CGU arginine	U
	CUC leucine	CCC proline	CAC histidine	CGC arginine	C
	CUA leucine	CCA proline	CAA glutamine	CGA arginine	A
	CUG leucine	CCG proline	CAG glutamine	CGG arginine	G
A	AUU isoleucine	ACU threonine	AAU asparagine	AGU serine	U
	AUC isoleucine	ACC threonine	AAC asparagine	AGC serine	C
	AUA isoleucine	ACA threonine	AAA lysine	AGA arginine	A
	AUG(start) methionine	ACG threonine	AAG lysine	AGG arginine	G
G	GUU valine	GCU alanine	GAU aspartate	GGU glycine	U
	GUC valine	GCC alanine	GAC aspartate	GGC glycine	C
	GUA valine	GCA alanine	GAA glutamate	GGA glycine	A
	GUG valine	GCG alanine	GAG glutamate	GGG glycine	G

Figure 8.2 Codon and Amino Acid Table

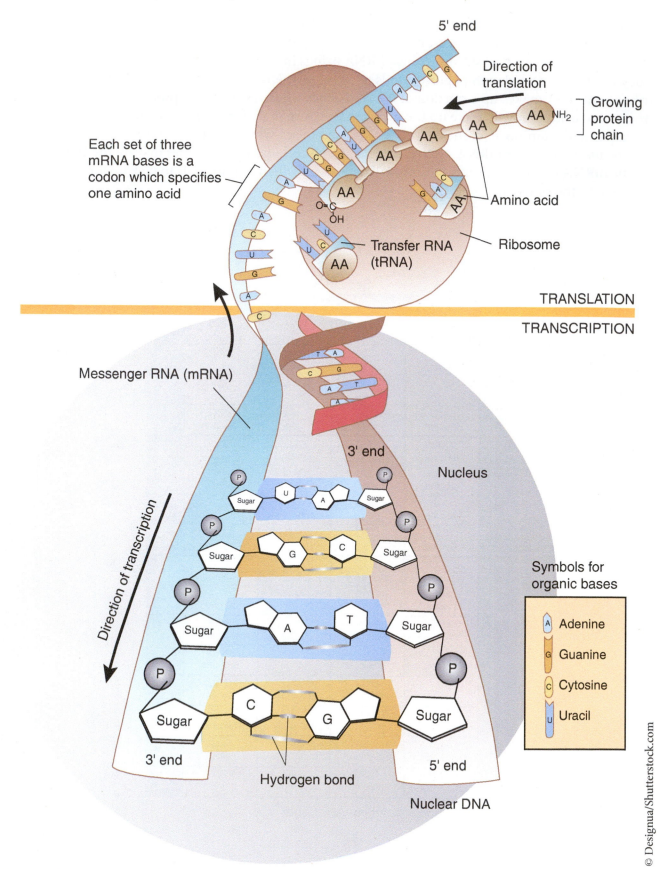

5' end

Direction of translation

Growing protein chain

AA NH₂

Each set of three mRNA bases is a codon which specifies one amino acid

AA

AA

AA

AA

AA

AA

Amino acid

AA

O=C
OH

Transfer RNA (tRNA)

Ribosome

AA

TRANSLATION

TRANSCRIPTION

Messenger RNA (mRNA)

3' end

Nucleus

Direction of transcription

P

Sugar   U   A   Sugar   P

P

Sugar   G   C   Sugar   P

P

Sugar   A   T   Sugar   P

P

Sugar   C   G   Sugar   P

Symbols for organic bases

A   Adenine

G   Guanine

C   Cytosine

U   Uracil

3' end

Hydrogen bond

5' end

Nuclear DNA

Figure 8.3 Comparison of DNA and RNA

Work the following:

DNA anti-sense	DNA sense	mRNA	tRNA	amino acids coded for
A	_____	_____	_____	
T	_____	_____	_____	
G	_____	_____	_____	_____
G	_____	_____	_____	
A	_____	_____	_____	
C	_____	_____	_____	_____
C	_____	_____	_____	
G	_____	_____	_____	
C	_____	_____	_____	_____
T	_____	_____	_____	
T	_____	_____	_____	
C	_____	_____	_____	_____

LAB 9

Mitosis

Cell division is central to the life of all organisms. All cells come from pre-existing cells as a result of cell division. If a multicellular organism is to continue to live, it must create new cells at a rate as fast as that at which its cells die. For example, in adult humans millions of cells must divide every second simply to maintain the status quo.

Eukaryotic cells may experience two types of cell division: mitotic cell division and meiotic cell division. This exercise deals with mitotic cell division. Prokaryotic cells divide by binary fission. They do not divide by either mitotic or meiotic cell division.

Mitotic cell division consists of two sequential processes: nuclear division (called **mitosis**) and cytoplasmic division (called **cytokinesis**). Mitotic cell division produces daughter cells that are genetically identical to each other and also genetically identical to the cell which divided to give rise to them.

During most of its lifetime, a given cell is not dividing. All cells have what is called a **cell cycle**, and the cell division phase accounts for only a small fraction of the total cell cycle. The phase of the cell cycle when a cell is not dividing is denoted as **interphase**. A typical cell cycle for eukaryotic cells is presented on the following page.

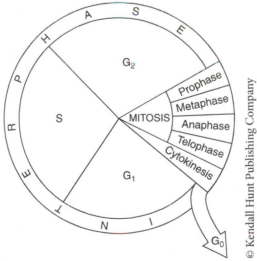

Figure 9.1 The Eukaryotic Cell Cycle.

Adapted from *BI102 Laboratory Manual: General Cellular Biology* by Washburn University. Copyright © 2008 by Kendall Hunt Publishing Company. Reprinted by permission.

$M =$ M phase = period of mitosis and cytokinesis.

$G_1 =$ First gap phase = interval between the end of mitotic cell division and the start of DNA synthesis. (Regulation of the length of the cell cycle occurs primarily by arresting at a specific point of G_1. Therefore, this phase has the most variable length.)

$S =$ S phase = specific part of interphase during which DNA synthesis occurs. (By the end of this phase, each chromosome consists of two identical chromatids.)

$G_2 =$ Second gap phase = interval between the end of DNA synthesis and the start of mitosis.

$G_0 =$ A temporary or permanent exit from the cell cycle.

(Cells often pause in G_1 before DNA replication and enter this state [G_0], where they may remain for days to years. Some cells never reenter the cell cycle. At any one time, most of the cells of a multicellular organism are in the G_0 phase.)

G_1, S, and G_2 are subdivisions of the portion of the cell cycle referred to as **interphase**. Interphase normally comprises 90% or more of the total cell cycle time.

The M phase of the cell cycle begins with mitosis and ends with cytokinesis. Mitosis is subdivided into four phases: prophase, metaphase, anaphase, telophase.

During **prophase** the following events occur:

1. The chromatin, which is diffuse in interphase, slowly condenses into well-defined chromosomes.

    (Each chromosome has duplicated during the preceding S phase and consists of two sister chromatids, joined at a specific point along their length by a region known as the **centromere**.)
2. The nucleolus begins to disassemble and gradually disappears.
3. In animal cells, paired centrosomes (each containing paired centrioles) move to opposite poles. Centrosomes are microtubule organizing centers (MTOC).

    (The cell's original centrosome replicates by a process that begins just prior to the S phase and continues through G_2 to give rise to two centrosomes. Each centrosome has a group of microtubules radiating from it. Collectively, these microtubules form the aster. Higher plant cells lack centrioles and asters.)
4. The **mitotic spindle** begins to develop outside the nucleus.

    (The spindle consists of fibers constructed of microtubules and microtubule-associated proteins. Cytoskeletal microtubules disassemble, and their tubulin dimers start to reassemble to form the mitotic spindle. Each spindle fiber is, therefore, a cluster of microtubules. The first fibers to form are **polar fibers**, which extend from the two poles of the spindle toward the equator of the cell.)
5. The nuclear envelope breaks into membranous fragments indistinguishable from ER.
6. Specialized structures called **kinetochores** (protein complexes) develop on either face of the centromeres and become attached to a special set of microtubules called **kinetochore fibers**. These fibers radiate in opposite directions from each side of each chromosome.

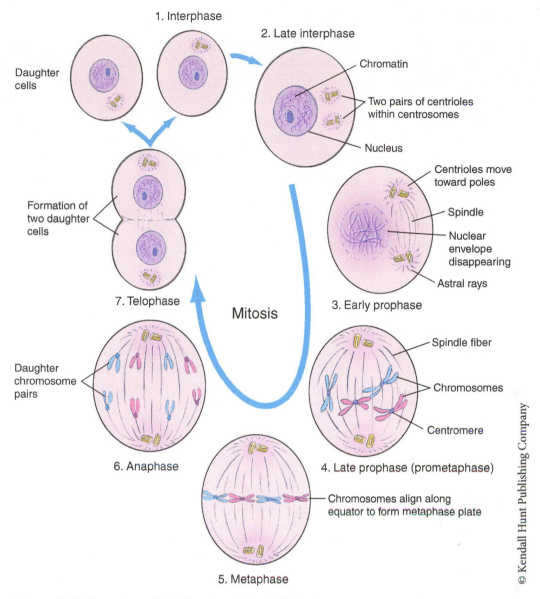

Figure 9.2 Overview of the Stages of the Cell Cycle

During **metaphase**, the following events occur:

1. The chromosomes become arranged so that their centromeres all lie in one plane (the **metaphase plate**) near the middle of the cell.

   (The kinetochore fibers seem to be responsible for aligning the chromosomes halfway between the spindle poles and for orienting them with their long axes at right angles to the spindle axis.)

During anaphase, the following events occur:

1. Motor molecules associated with each kinetochore separate the sister chromatids of each chromosome.
2. Each chromatid (now called a chromosome) is moved slowly toward a spindle pole by the motor molecules of the kinetochore as they "walk" along the kinetochore microtubules.
3. Kinetochore fibers (microtubules) progressively shorten (by depolymerizing at their kinetochore ends) as the chromosomes approach the poles.
4. Polar fibers (microtubules) elongate and move the two poles of the spindle further apart. These nonkinetochore microtubules elongate the whole cell along the polar axis.

During **telophase**, the following events occur:

1. As the separated chromosomes (formerly called sister chromatids) arrive at the poles, the kinetochore fibers disappear.
2. A new nuclear envelope reforms around each group of chromosomes.
3. The condensed chromatin disperses once more.
4. Nucleoli reappear.

In animal cells, cytokinesis is accomplished by the development of a **cleavage furrow**. The cleavage furrow forms as a result of the interaction of actin and myosin filaments, which form a contractile ring just below the plasma membrane. In plant cells, cytokinesis is accomplished by the formation of a **cell plate**, rather than a cleavage furrow. The cell plate is assembled as Golgi-derived vesicles containing cell wall material are directed to the center of the dividing cell by microtubules. The vesicles then fuse with each other to create the cell plate.

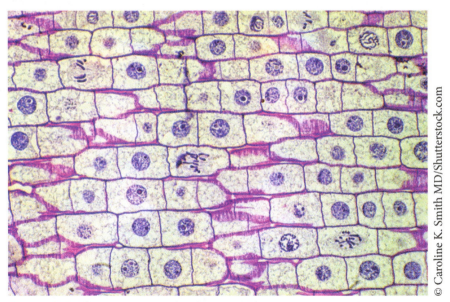

© Caroline K. Smith MD/Shutterstock.com

Figure 9.3 Photomicrograph of onion (*Allium*) root tip mitosis

Onion Root Tip Mitosis

Obtain a prepared slide of an onion (*Allium*) root tip. Focus on the slide so that you are viewing an area just above the tip. Do not focus on the very tip because these are protective root cap cells and are not undergoing mitosis. Using proper focusing procedures, work your way up to high power to observe individual cells. Refer to the description of each phase to help you locate and identify each phase.

For all sketches, label the total magnification of the microscope view.

1. Locate a cell in <u>interphase</u>. Sketch this cell and label the nuclear envelope, nucleolus, and cell wall.

2. Locate a cell in <u>prophase</u>. Can you distinguish the chromosomes?

   Is a distinct nuclear envelope obvious?

   Sketch this prophase cell, indicate the direction of the spindle fibers, and label the chromosomes, centromeres, and poles.

3. Locate a cell in <u>metaphase</u>. How can the metaphase cell be distinguished from prophase?

   Sketch the metaphase cell and label the chromosomes, spindle fibers, and poles.

4. Locate a cell in <u>anaphase</u>. What is the major event of anaphase?

   Are two groups of chromosomes visible?

Sketch and label the anaphase cell.

5. Locate a cell in <u>telophase</u>. Is the cell plate visible?

Has the nucleus reformed?

Sketch and label the telophase cell.

6. If onions have a chromosome number of 16, how many chromosomes are in each of the daughter cells formed by cytokinesis at the end of telophase?_____ How many chromosomes were in the original cell?_____

Phases of the Cell Cycle in the Onion Root Tip

Since you are familiar with the events of mitosis and the stages of the cell cycle, you can now estimate the relative duration of each phase of the cell cycle. The duration of each phase is estimated by recording the frequency with which you find each phase in regions where cell division is actively taking place. The frequency of a phase is an indication of the relative length of that phase.

1. Obtain a prepared slide of an onion root tip.
2. Under high pressure, examine a single field of view (the area visible in the microscope with the slide stationary) and count the number of cells in the phases of the cell cycle. Make sure you are surveying the actively dividing area of the onion root tip. Repeat this count in at least two more non-overlapping fields.

TABLE 9.1 Percentage of Cells in Each Phase of the Cell Cycle.

Number of Cells	Field 1	Field 2	Field 3	Total	Percent of Grand Total (Total/Grand Total × 100)
Interphase					
Prophase					
Metaphase					
Anaphase					
Telophase					
Grand Total					

The duration of mitosis varies for different tissues in the onion. However, prophase is always the longest phase (1–2 hours), and anaphase is always the shortest (2–10 minutes). Metaphase (5–15 minutes) and telophase (10–30 minutes) are also relatively short in duration. Interphase may range from 12–30 hours.

Consider that it takes, on average, 16 hours (960 minutes) for onion root tip cells to complete the cell cycle. You can calculate the amount of time spent in each phase of the cell cycle from the percentage of cells in that stage:

Percentage of cells in stage × 960 minutes = minutes of cell cycle spent in stage.

3. Calculate the following (convert the times to hours and minutes):

   Time spent in: prophase   _____ hr. _____ min.

   metaphase   _____ hr. _____ min.

   anaphase   _____ hr. _____ min.

   telophase   _____ hr. _____ min.

   Total time spent in mitosis   _____ hr. _____ min.

   Total time spent in interphase   _____ hr. _____ min.

   What percentage of the cell cycle is spent in mitosis?

$$\frac{\text{total time spent in mitosis (in min.)}}{960 \text{ min.}} =$$

What percentage of the cell cycle is spent in interphase?

100% − % of cell cycle spent in mitosis =

How do your results compare with what is generally known about the onion root tip cell cycle?

Whitefish Mitosis

Obtain a slide labeled <u>whitefish</u> <u>blastula</u> to study mitotic cell division in an animal. The blastula is a stage of embryonic development. When a sperm fertilizes an egg cell, a zygote is produced. This zygote divides by mitotic cell division to produce a ball of cells known as a blastula. Mitotic cell divisions will be one of the major events that will allow this early embryonic stage to grow and develop into an adult organism.

Use proper focusing procedures to work your way up to high power. Locate and examine a cell in each phase of the cell cycle. Make a sketch of each phase, including total magnification of the microscope view.

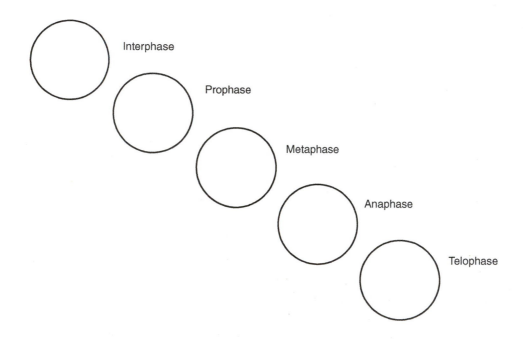

If the chromosome number of the whitefish cell were 24, how many chromosomes would be in each daughter cell?_____.

Indicate three places in your body where there is a high rate of mitotic cell division.

a. _____

b. _____

c. _____

Models of Mitotic Cell Division

A series of models that shows mitotic cell division as it occurs in animal cells is available in the laboratory. Study these so that you understand what is being represented by each model.

LAB 10

Bacteriology Preparation

Bacteria are prokaryotic, unicellular organisms. They may be producers, consumers, or decomposers. Bacteria are traditionally classified based on shape, mode of nutrition, motility, and staining properties. DNA technology has revolutionized how bacteria are classified in a phylogenetic sense, but the traditional ways of classifying these organisms still have many practical applications, particularly in medical settings.

The following instructions are for groups of two students to produce a Petri dish with nutrient agar in order to inoculate for use in the next lab.

1. Boil one test tube full of agar per Petri dish used. Allow the test tubes to slow boil for 10 minutes.
2. Obtain a clean Petri dish and pour the agar into the bottom half up to about 6mm in depth or at least to the point where the bottom of each side is covered. Take care not to burn yourself. You will need to gently swirl the agar in the bottom of the plate to get full coverage.
3. Place the top of Petri dish over most of the plate (leave the top open a slight bit for the escape of water vapor). Let stand until the agar is congealed (25 to 30 minutes).
4. Using a clean cotton swab, sample a surface or material of your choice (no bodily fluids or toilet seats) and inoculate the plate by streaking with the swab as shown by your instructor. Take care not to penetrate the surface.
5. Replace lid, invert Petri dish, label it with your names and a brief description of the item sampled (on the edge of the dish, NOT the lid), and store it as your instructor directs.

LAB 11

Use and Construction of a Taxonomic Key

A taxonomic key is used to identify specimens or items. The type of key described in this exercise is a dichotomous key, from the Greek meaning "divided into two parts." It is a key that uses a series of statements arranged in pairs or couplets, such as 1a and 1b; 2a and 2b; and so on. Each couplet includes one or more characteristics, and they are structured such that one of the statements characterizes the specimen to be identified, while the other statement does not apply. Both statements of the couplet are read and a decision is made as to what statement characterizes the specimen. That statement directs the user to another couplet or identifies the specimen. For example:

1a 2
1b ....Specimen identified

If the specimen is characterized by statement 1a, the user continues to couplet 2; if the specimen is characterized by 1b, the specimen is identified.

The following are some suggestions for constructing a dichotomous taxonomic key:

1. Use characteristics that are constant and do not change (i.e., do not use a characteristic that may be seasonal).
2. If measurements are being used, be exact. Use a number rather than terms like small or large.
3. Use characteristics that can be seen on the specimen and not assumed.
4. If possible, start each couplet with the same word.
5. The descriptive statements should be written in couplets with each statement ending with the number of the next couplet to be read or with the name of the specimen.

The following are two examples of a short simple dichotomous key used to identify the same five fruits shown in the diagram below:

Photo courtesy Jimmy Nevins

1a. Shape is not elongated or round 2

1b. Shape is elongated ....Banana

2a. Outer covering or skin is thin (<1 mm) 3

2b. Outer covering or skin is thick (>1 mm) ....Orange

3a. Grows singly 4

3b. Grows in clusters ....Grape

4a. Small stem projects from top of the fruit ....Apple

4b. No stem projects from top of the fruit ....Blueberry

In some keys the first couplet may divide the specimens into two groups. For example:

1a. Outer covering is thick (>1 mm) 2

(Characteristic of the banana and orange.)

1b. Outer covering is thin (<1 mm) 3

(Characteristic of the apple, grape and blueberry)

2a. Round shape ....Orange

2b. Elongated shape ....Banana

3a. Round and grows singly 4

3b. Round and grows in cluster ....Grape

4a. A small stem projects from the top of the fruit ....Apple

4b. No stem projects from the top of the fruit ....Blueberry

USING A TAXONOMIC KEY

1. If available, your instructor will provide samples of branches from common plants native to the area along with a taxonomic key.
2. Following the key, identify each of the samples and verify the results with the instructor.

Photo courtesy Jimmy Nevins

Specimen	Names

CONSTRUCTING A TAXONOMIC KEY

The directions may be altered by the instructor.

1. Use the assigned specimens to construct a taxonomic key. Depending on the specimens, mount pictures of the specimens on a sheet of paper or submit the actual specimens, as appropriate.
   a. The descriptive statements should be written in couplets with each statement ending with the number of the next couplet to be read or with the name of the specimen.
   b. All characteristics used in the key should be observed on the specimen and not assumed to be present. If color is used as a characteristic, it must be seen in the picture.
2. Type the key and attach pictures of the specimens or the specimens, as appropriate.

LAB 12

Bacterial Smear Preparation and Simple Staining

SIMPLE STAINING

Because bacteria are very small and almost completely transparent, they are very difficult to observe without staining. The use of a single stain to color bacterial cells is called simple staining. The outer surface of most bacteria is negatively charged, so positively charged molecules are attracted to, and bind with, the bacterial cell surface. In general, basic (positively charged) dyes are used as direct simple stains. Stains color the bacterial cells to create contrast between the bacteria and the background so that the cells are clearly visible with the microscope. With simple staining, cell shape, size, and arrangement can be observed. Stained bacteria can be measured for size and are classified by their shapes and groupings.

BACTERIAL SMEAR PREPARATION

The first step in the staining process is to prepare a smear. A smear is a dried preparation of bacterial cells on a glass slide. The purpose of making a smear is to adhere/fix the bacteria to the slide to prevent the sample from being lost during the staining process. Smears can be prepared from liquid cultures or from cultures grown on an agar medium.

a. When using a liquid culture, one loopful of culture is smeared onto a glass slide and allowed to air dry. The cells in the dried smear are then "fixed" to the slide by briefly heating. This process is known as heat fixation.

b. When using growth from an agar medium, a loopful of water is placed on the slide, and a very small amount of culture is mixed with the water to separate and suspend the bacteria. The suspension is then spread out, air dried, and heat fixed. In a good smear, the bacteria are evenly spread out on the slide and individual organisms are visible microscopically.

BACTERIAL CELL MORPHOLOGY

Bacteria have rigid cell walls that function to maintain a constant shape. There are three basic shapes: cocci, bacilli, and spiral. Bacterial cells group together as they multiply, and the arrangement of these groups is often characteristic of a genus or species (see **Figure 12.1**).

Adapted from *Fundamentals of Microbiology for Allied Health* by H. Kathleen Dannelly, Angela K. Chamberlain, and William M. Chamberlain. Copyright © 2009 by Kendall Hunt Publishing Company. Reprinted by permission.

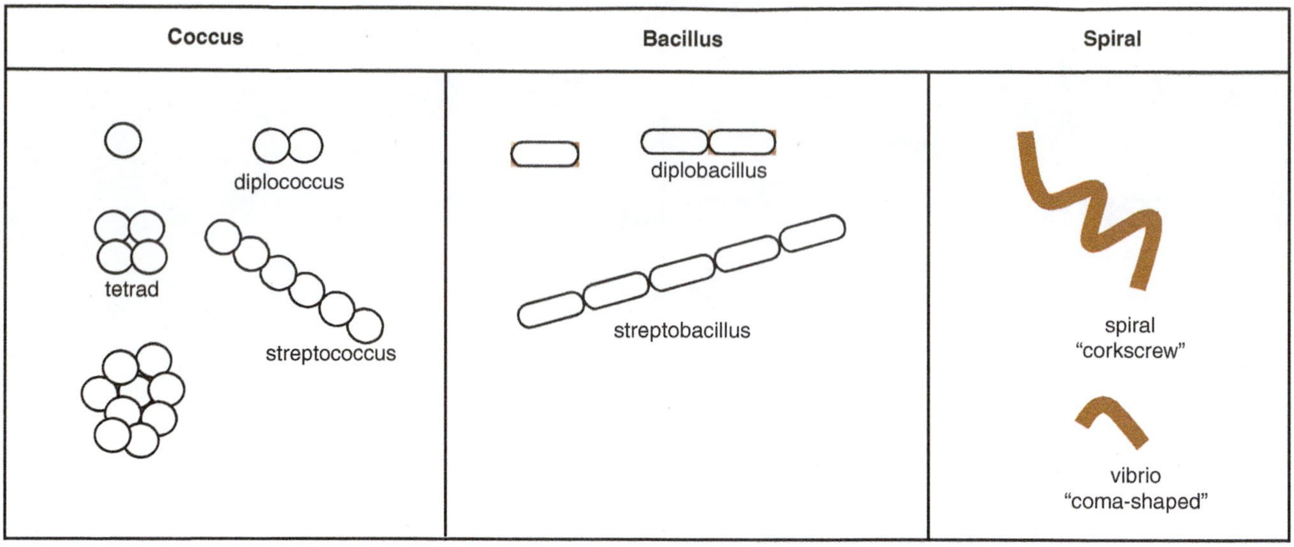

Coccus	Bacillus	Spiral

diplococcus

tetrad

streptococcus

diplobacillus

streptobacillus

spiral "corkscrew"

vibrio "coma-shaped"

© Kendall Hunt Publishing Company

(a) Cocci

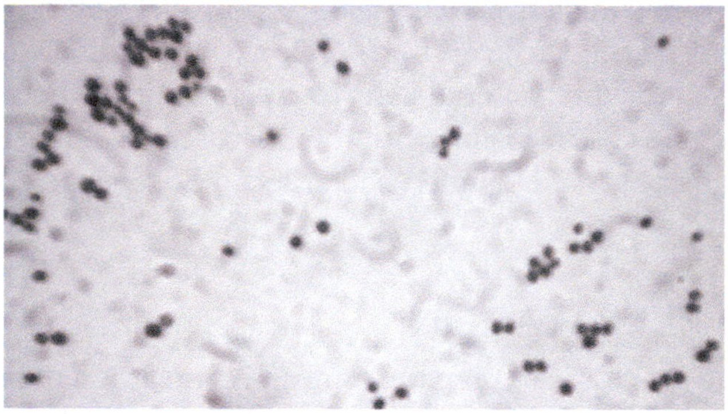

(c) Spiral

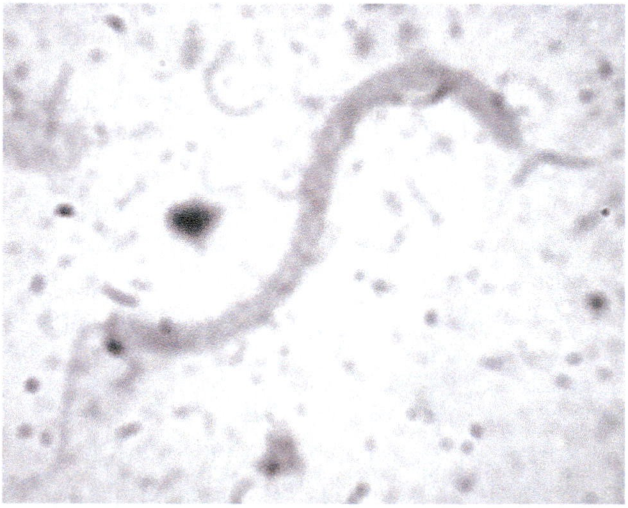

(a) Bacilli

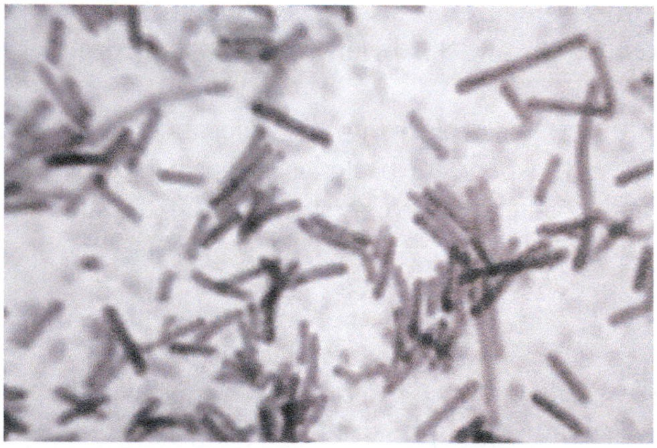

© Kendall Hunt Publishing Company

Figure 12.1 Common shapes and arrangements of bacterial cells.

1. Place a small drop of water on a clean microscope slide*.

2. With the inoculating loop, aseptically addbacterial culture to the drop of water.

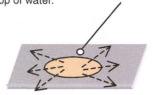

3. Mix the bacteria with the water to a fine suspension, and spread it out.

4. Allow the suspension to air dry.

5. Heat fix the suspension by placing the side on a hot plate set to "low" on "2".

Figure 12.2 Bacterial smear preparation.

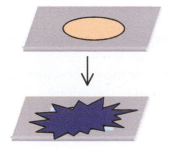

1. Place slide on staining rack and Cover prepared smear with stain.

2. Let sit 1 minute.

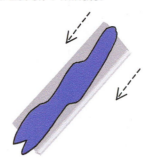

3. Rinse slide thoroughly with water. Use caution with pressurized hoses.

4. Gently blot dry with paper towels; do not rub.

5. Use proper focusing procedures to get to 400x magnifiaction.

6. Call instruction over for assistance with oil immersion techniques.

Figure 12.3 Procedure for simple staining of bacterial smears.

PROCEDURE

1. Choose an isolated/pure colony on one of the plates you obtained by sampling the environment.
2. Use a stereoscope to view and describe the colony. Colony morphology information is available in the laboratory.
3. Follow the procedure diagrammed in **Figure 12.2,** to prepare a smear using the isolated colony as the source.
4. Follow the procedure diagrammed in **Figure 12.3.**
5. Observe the material on the slide and report your results.
6. Use *Bacillus cereus* as the test organism, and repeat Steps 1-4 above.

*Some students find it helpful to draw a circle with a wax pencil on the underside of the slide to mark the place where they are going to make the smear.

RESULTS

For each of the test organisms, answer the following questions:

1. For each organism, use the stereoscope and colony morphology information to describe the form, elevation, and margin of the colony.

   Environmental Sample:

           Form:

           Elevation:

           Margin:

   Bacillus cereus:

           Form:

           Elevation:

           Margin:

2. For each organism, describe what your stained smear looks like from the macroscopic view (to the unaided eye). Is it thick and dark blue? Or is it thinned out and faint blue?

   Environmental Sample:

   B. cereus:

3. What do you see when you first view your smear with low power? Describe it.

   Environmental Sample:

B. cereus:

4. Describe what you see when you view your smear on high dry.

Environmental Sample:

B. cereus:

5. Find a field of view where the cells appear to be in a single layer. Draw what you see when you view your slide through the oil immersion lens.

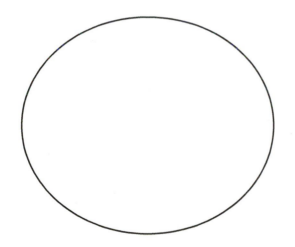

Environmental Sample

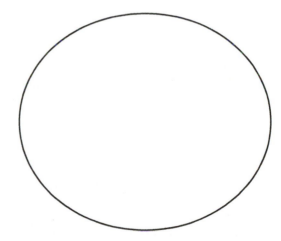

Bacillus cereus

6. Are there areas on the slide where the cells are heaped up into multiple layers? (This is the most common mistake made by beginning microbiology students.) You may notice that in these areas, the cells did not stain properly and there are very large dark blue areas where no cells can be discerned under the scope. Next time you prepare a stain, take care not to get the smear too thick.

7. What is the purpose of staining bacterial cells?

8. What is the purpose of heat fixing a bacterial smear?

9. What are the characteristics of a good bacterial smear?

LAB 13-A

Protists & Fungi

PROTIST KINGDOM

The organisms included in the Protist Kingdom are diverse, sharing few characteristics and lacking a close evolutionary ancestry. All Protist are eukaryotic and many are found in watery environments. They can be uni- or multi-cellular, symbiotic or parasitic, and may or may not perform photosynthesis. The term, Protista is not a true kingdom, but a catchall category for organisms that are not animals, plants, fungi, or eubacteria. Protists were commonly divided into "animal-like" **protozoa**, "plant-like" **algae,** and "fungus-like" slime- and water-molds. Organisms once categorized as protists are currently undergoing genetic evaluation to more accurately determine their phylogeny.

© Kendall Hunt Publishing Company.

Figure 13.1 The green lichen seen here is actually a symbiotic mixture of protist and fungus. The protist is an algae and provides nutrients through photosynthesis while the fungus wraps around the algae and helps to secure it to the trunk of the tree. Without each member of the team working together, the lichen would not survive.

The most commonly studied protist is the **Amoeba**. *Amoebas* are single celled organisms that reproduce asexually by mitosis. Their shape is constantly changing because of their arm-like extensions called **pseudopods.** Aside from helping the amoeba change shape, the pseudopods also help the *amoeba*

Adapted from From *BIO 105: Introduction to Biology Lab Manual* by Stephanie M. Brown. Copyright © 2009 by Kendall Hunt Publishing Company. Reprinted by permission.

move and even eat! The pseudopods will surround the amoeba's prey to trap it and use phagocytosis to ingest it. *Amoebas* are known to eat plankton, diatoms, and *Paramecium*. If we have live amoeba available in the lab, we will try to observe this behavior.

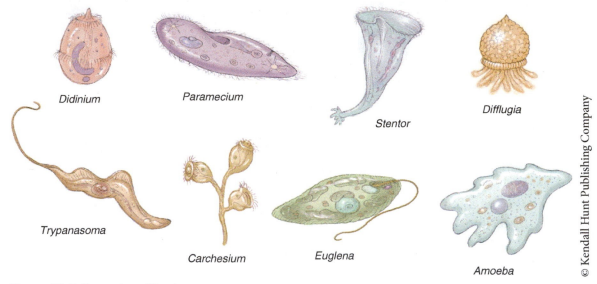

Figure 13.2 Examples of Protista

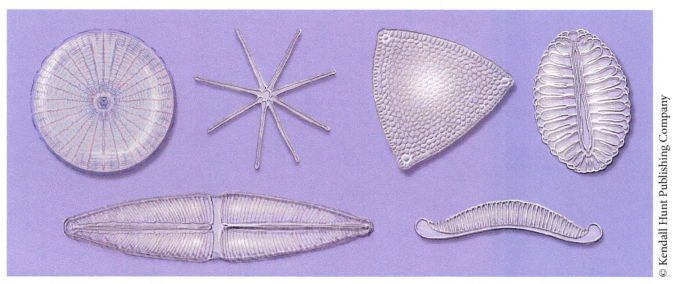

Figure 13.3 Examples of Diatoms

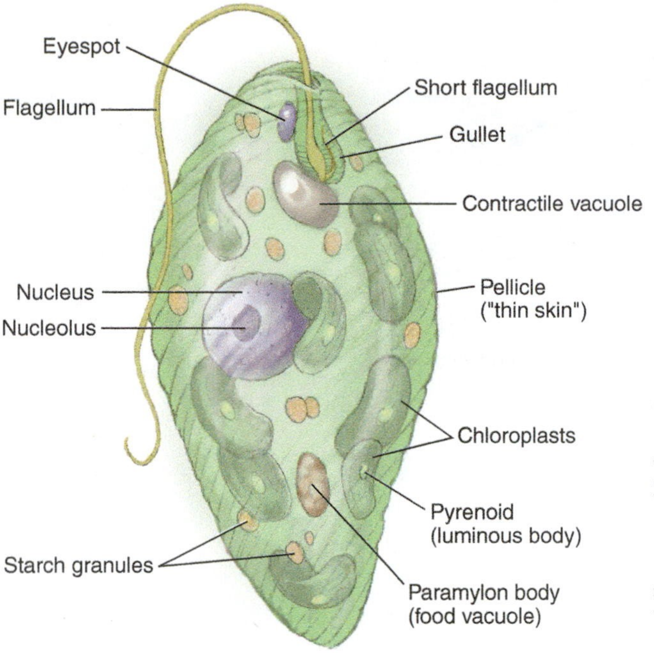

Eyespot

Flagellum

Short flagellum

Gullet

Contractile vacuole

Nucleus

Nucleolus

Pellicle
("thin skin")

Chloroplasts

Pyrenoid
(luminous body)

Starch granules

Paramylon body
(food vacuole)

© Kendall Hunt Publishing Company

Figure 13.4 *Euglena*

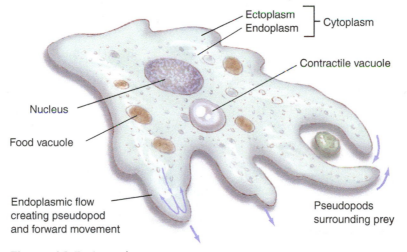

Ectoplasm
Endoplasm

Cytoplasm

Contractile vacuole

Nucleus

Food vacuole

Endoplasmic flow
creating pseudopod
and forward movement

Pseudopods
surrounding prey

© Kendall Hunt Publishing Company

Figure 13.5 *Amoeba*

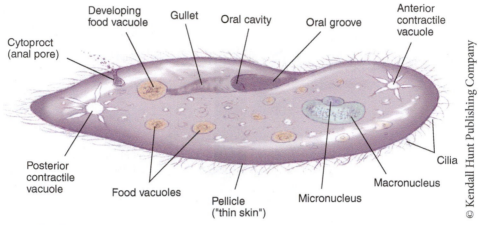

Cytoproct (anal pore)

Developing food vacuole

Gullet

Oral cavity

Oral groove

Anterior contractile vacuole

Posterior contractile vacuole

Food vacuoles

Pellicle ("thin skin")

Micronucleus

Macronucleus

Cilia

© Kendall Hunt Publishing Company

Figure 13.6 *Paramecium*

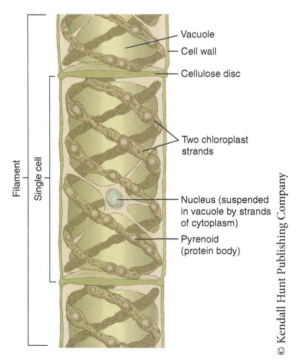

Vacuole

Cell wall

Cellulose disc

Two chloroplast strands

Nucleus (suspended in vacuole by strands of cytoplasm)

Pyrenoid (protein body)

Filament

Single cell

© Kendall Hunt Publishing Company

Figure 13.7 *Spirogyra*

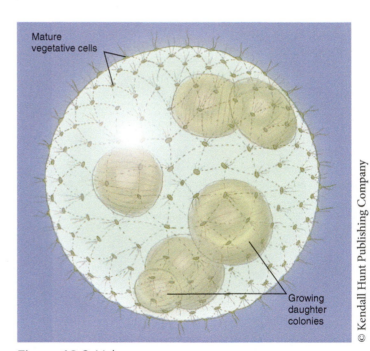

Mature vegetative cells

Growing daughter colonies

© Kendall Hunt Publishing Company

Figure 13.8 Volvox

Focus each of the prepared slides starting with the 4X objective and continuing through to the 40X objective. Draw your specimens under each objective below and label the recognizable structures.

Amoeba _____

40X	100X	400X

Paramecium _____

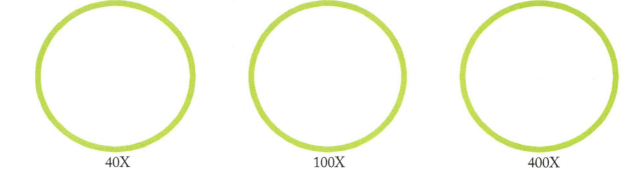

40X	100X	400X

Euglena _____

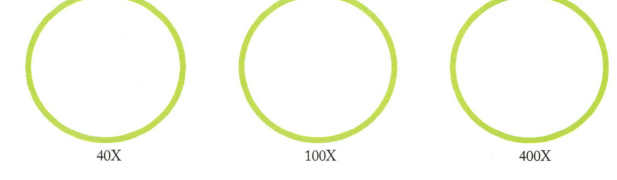

40X	100X	400X

Volvox _____

40X	100X	400X

QUESTIONS

1. What is the dark dot in the center of the *Amoeba*?

   _____

2. What is the function of the *Paramecium*'s cilia?

   _____

3. What is the function of the *Euglena*'s flagella?

   _____

4. How can you differentiate between the *Paramecium* and the *Euglena*?

   _____

5. How is the *Volvox* different from the other Protists you have observed?

   _____

6. How did the prepared slides compare to the live samples?

   _____

GOGGLES AND GLOVES MUST BE WORN

Obtain a drop or two of the **live** *Paramecium* culture with a clean pipete from the bottom of the specimen sample jar. Take care not to disrupt the sample jar too much. Add a drop of methylcellulose to the organisms to slow down their movement. Add a drop of yeast stained with Congo red dye. Stir gently with a toothpick to mix. Slowly add a coverslip to the top of the wet mount and observe.

Congo red is a dye that is red in an acidic pH and blue in a basic pH. The yeast will be red and the paramecium will look as if they are bumping into the yeast. Hopefully you will see the paramecium eat the yeast and the color of the yeast will change from red to blue as the pH changes as the paramecium digests it. You may also see the cilia moving around the *Paramecium* as noticed by the wave-like movement of the yeast away from the *Paramecium*.

Paramecium eating yeast stained with Congo red dye

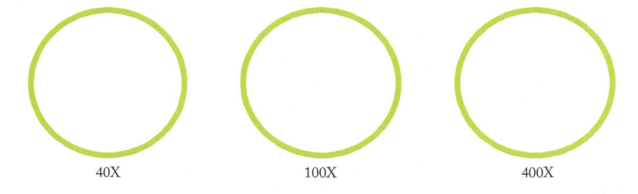

40X   100X   400X

If a specimen is not already prepared, obtain a drop or two of the *live* Amoeba culture with a clean pipette from the bottom of the specimen sample jar. Take care not to disrupt the sample jar too much. The *Amoeba* usually stick to the bottom of the jar. Add a drop of the live *Paramecium* culture directly on top of the *Amoeba* drop. Slowly add a coverslip to the top of the wet mount and observe.

Amoeba eating *Paramecium*

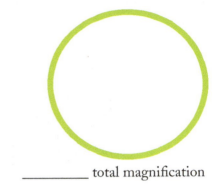

_____ total magnification

Add a drop of methylene blue dye to the *Amoeba* and *Paramecium* slide from above by placing the small drop of dye next to the edge of the coverslip on your slide and wicking the dye underneath to stain the cells. Observe. Note the change in the cell's behavior.

The blue dye will initiate the release of spear-like trichocysts by the *Paramecium*; this is a defense mechanism. Try to observe the cells as the methylene blue is approaching them. Of course the dye will also stain the nucleus of the cells as well.

Amoeba eating *Paramecium* with methylene blue dye_____

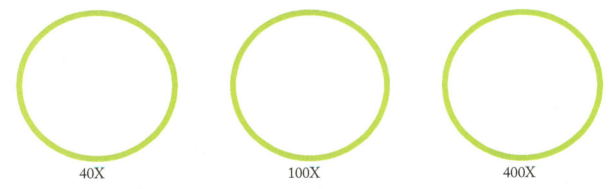

40X   100X   400X

QUESTIONS

7. Were you able to observe any eating from either organism?

_____

8. How did the previously viewed prepared stained slides compare to the live samples? Which did you prefer and why?

_____

9. How did the *Amoeba* and *Paramecium*'s behavior change once you added the methylene blue dye?

_____

FUNGI

We have all seen fungus. On bread. On fruit. On the ground or a tree. Even on your feet! Fungi can grow on pretty much anything. One common fungus used to make bread rise is the yeast *Saccharomyces cerviseae*. Another type of yeast *Candida albicans* grows in your mouth. Aside from yeast, mold is also a type of fungus. We will look at three different types of mold under the microscope: *Penicillium*, *Aspergillus*, and *Rhizopus*.

© Patricia Chumillas/Shutterstock.com

Figure 13.9 Mold on a slice of bread

© Christopher Meade/Shutterstock.com

Figure 13.10 The mold *Penicillium* from which the antibiotic penicillin is produced.

Figure 13.11 Clockwise from back left: Mushrooms: *Coprinus* (note the black liquid from the gills), *Morchella* (large and wrinkled caps) and cup fungus (catches rain for spore dispersal)

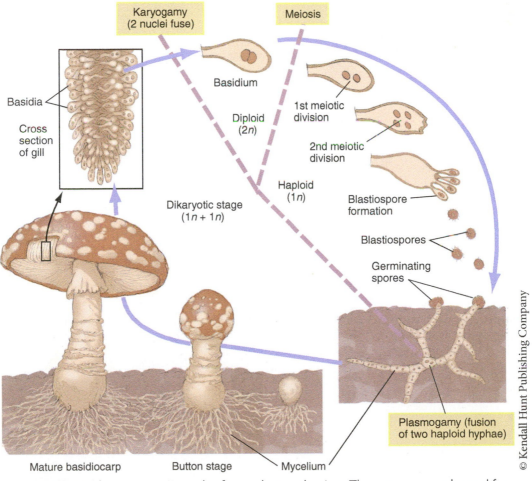

Figure 13.12 Mushrooms are a result of sexual reproduction. The spores are released from club shaped basidia found in the gills.

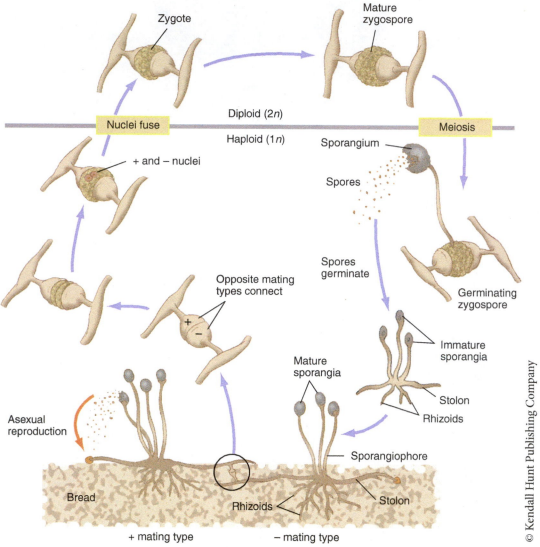

Zygote

Mature
zygospore

Nuclei fuse

Diploid (2*n*)

Meiosis

Haploid (1*n*)

Sporangium

+ and − nuclei

Spores

Spores
germinate

Germinating
zygospore

Opposite mating
types connect

+

−

Immature
sporangia

Mature
sporangia

Stolon

Rhizoids

Asexual
reproduction

Sporangiophore

Bread

Stolon

Rhizoids

+ mating type

− mating type

© Kendall Hunt Publishing Company

Figure 13.13 *Rhizopus*, black bread mold has two modes of reproduction. It can form zygospores through a form of sexual reproduction and it can produce and release spores from the mature sporangia in asexual reproduction.

PROCEDURE: Focus each of the prepared slides starting with the 4X objective and continuing through to the 40X objective. Draw your specimens under each objective below and label the recognizable structures.

Penicillium _____

40X   100X   400X

Aspergillus _____

40X   100X   400X

Rhizopus _____

40X   100X   400X

How do the spore producing structures differ between these three fungi?

_____

_____

_____

What is a hypha?

_____

_____

Use the space below to sketch any fungi samples available in the lab:

Use the taxonomic key, to identify each specimen to group and genus.

Group/Phylum	Genus (if given)
1.	
2.	
3.	
4.	
5.	
6.	
7.	
8.	
9.	
10.	

LAB 13-B

BIO 113: Kingdom Fungi

1a.	Non-photosynthetic	...go to 2
1b.	Photosynthetic	...go to 10
2a.	Hair-like, fuzzy or powdery	...go to 3
2b.	Not hair-like, large head structure, spherical	...go to 5
3a.	Sexual reproduction forms stalks and colored sporangium..............Rhizopus (Zygomycota)	
3b.	No sexual reproduction	...go to 4
4a.	Important in antibiotic production...................................Penicillium (Imperfect Fungi)	
4b.	Important in citric acid production...............................Aspergillus (Imperfect Fungi)	
5a.	All individual parts microscopic...Yeast (Unicellular Fungi)	
5b.	Some individual parts not microscopic	...go to 6
6a.	Reproductive structures umbrella-like or spherical	...go to 7
6b.	Reproductive structures must have stalk; cup-like or wrinkled head	...go to 9
7a.	Reproductive structure spherical with spore production inside.............Puffballs (Basidiomycota)	
7b.	Reproductive structure umbrella-like	...go to 8
8a.	Reproductive structure has stalk and gills underneath....................Mushroom (Basidiomycota)	
8b.	Reproductive structure is shelf-like....................................Shelf Fungi (Basidiomycota)	

9a.   Reproductive structure cup-like...……......
Cup Fungi (Ascomycota)

9b.   Reproductive structure with short stalk and wrinkled head........................
Morel (Ascomycota)

10a.   Completely green or with chlorophyll in distinct bodies.............
*Spyrogyra (Protista)

10b.   Parts are green to gray; crust-like or upright and branched...............…..
*Lichen (Algal Symbiosis)

                                             * not a Fungi /or some parts not Fungi

Specimen	Steps in classification	Name of Organism	Large Group (Phylum etc.)
1.			
2.			
3.			
4.			
5.			
6.			
7.			
8.			
9.			
10.			
11.			
12.			

LAB 14

Plants

Plants are multicellular, eukaryotic organisms that provide much of the oxygen we breathe. The characteristic structures of plant cells are the cell wall, central vacuole, and chloroplasts. A plant's chloroplasts contain the green pigment **chlorophyll**, which allows the plant to obtain its energy directly from the sun in the process of **photosynthesis.** A plant's life is a cycle of two phases: the **gametophyte** stage, during which the plant produces gametes; and the **sporophyte** stage, during which the plant produces spores. The repeated alternation between these two phases is known as **alternation of generations**. A plant is said to be **dominant gametophyte** if it spends most of its life in the gametophyte stage. Whereas a plant that spends most of its life in the sporophyte stage would be a **dominant sporophyte**. The four major groups of plants we will study are: moss, fern, gymnosperm, and angiosperm.

MOSS

Moss, or **bryophytes**, are a group of plants that do not have vascular tissue, seed, or flowers. They are typically found in shady, moist locations and grow very low to the ground, often resembling a soft green carpet. Moss require water for fertilization and the wind for spore dispersal. They are dominant gametophyte. The gametophyte is the green leaf-like portion of the moss. The sporophyte develops from the gametophyte and is dependent on the gametophyte for much of its nutrients. The sporophyte of the moss is composed of two parts: a stalk and a capsule containing spores. The capsule bursts to expel the spores that travel in the wind to their site of germination.

© Kendall Hunt Publishing Company.

Figure 14.1 Moss. Notice the tall sporophytes growing up from the shorter gametophytes

Adapted from *BIO 105: Introduction to Biology Lab Manual* by Stephanie M. Brown. Copyright © 2009 by Kendall Hunt Publishing Company. Reprinted by permission.

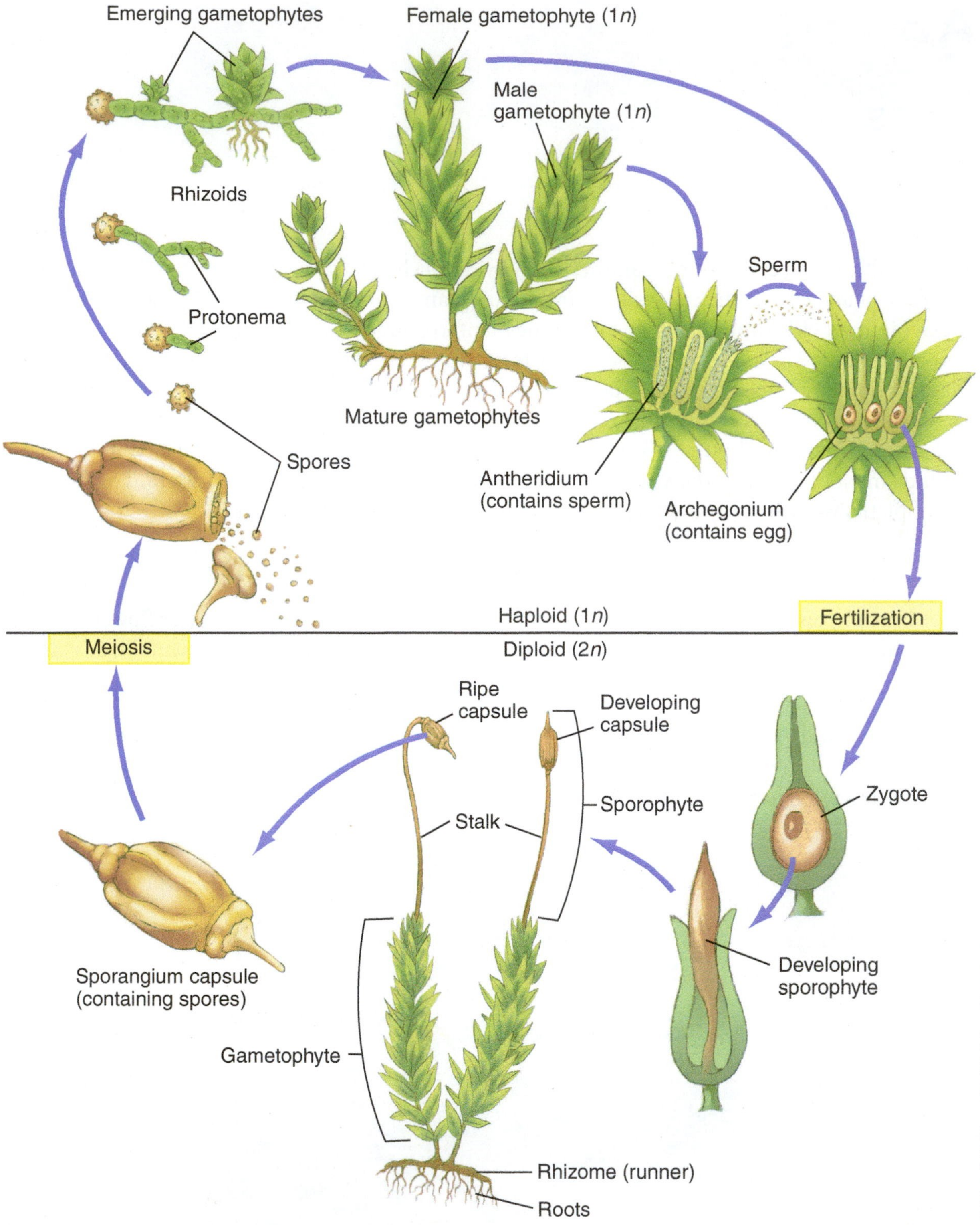

Figure 14.2 Moss Life Cycle

Obtain a slide from your instructor. Focus the slide starting with the 4X objective and then the 10X. Draw your specimen from the objective specified below.

Moss Capsule

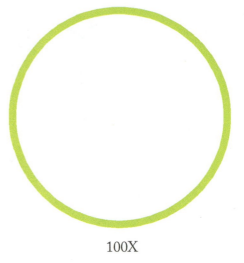

100X

QUESTION

1. What are the tiny red dots inside the moss capsule?

_____

FERN

Ferns, unlike moss, have vascular tissue. The vascular tissue is divided into two parts, the **xylem** and the **phloem**. The xylem transports water throughout the plant while the phloem transports the sugar produced through photosynthesis. Similar to mosses, ferns do not have seeds or flowers. Ferns are dominant sporophyte. In the sporophyte stage, a fern produces **sori** that resemble little dots on the underside of the fern frond. The sori are clusters of spores. In a fern's gametophyte stage, the fern is a small heart-shaped leaf or **prothallus**. Inside the prothallus, you can see many gametes. The male gamete is the **antheridium** and the female gamete is the **archegonium**.

© Kendall Hunt Publishing Company.

Figure 14.3 A fern frond on the left and a close up of the sori on the underside of a frond on the right.

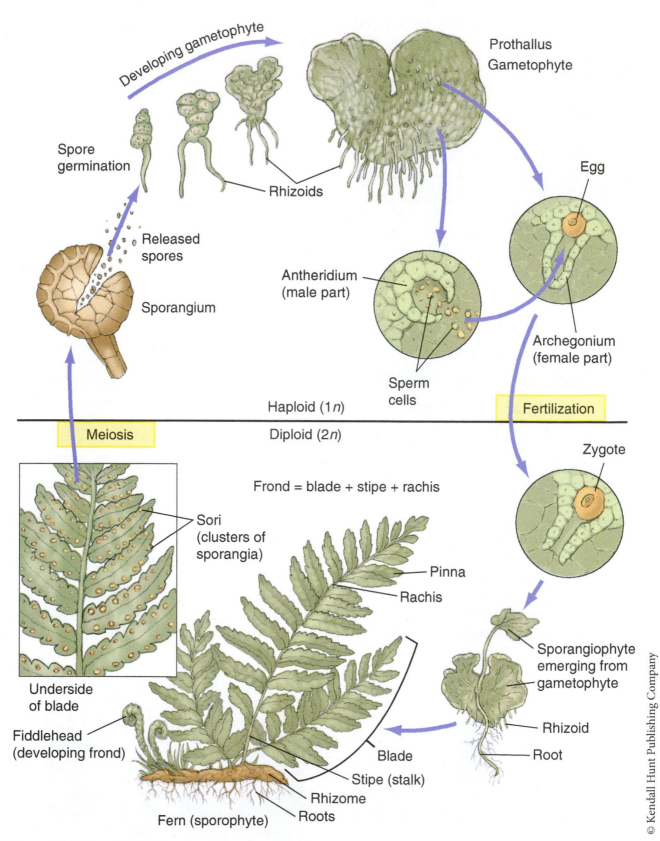

Developing gametophyte

Prothallus
Gametophyte

Spore
germination

Rhizoids

Egg

Released
spores

Antheridium
(male part)

Archegonium
(female part)

Sporangium

Sperm
cells

Haploid (1*n*)

Fertilization

Meiosis

Diploid (2*n*)

Zygote

Frond = blade + stipe + rachis

Sori
(clusters of
sporangia)

Pinna

Rachis

Sporangiophyte
emerging from
gametophyte

Underside
of blade

Rhizoid

Root

Fiddlehead
(developing frond)

Blade

Stipe (stalk)

Rhizome

Roots

Fern (sporophyte)

Figure 14.4 Fern Life Cycle

Obtain a slide from your instructor. Focus the slide starting with the 4X objective and moving through each objective up to the 40X objective. Draw your specimen under each objective below.

Prothallus

40X   100X

QUESTIONS

2. Looking at your drawing under the 4X objective, what shape does the prothallus resemble?

   _____

3. Label the antheridium in the 10X diagram above. Is this the male or female gamete?

   _____

4. Label the archegonium in the 10X diagram above. Is this the male or female gamete?

   _____

5. How are the spore producing structure of the moss and fern different?

   _____

GYMNOSPERM

Like ferns, gymnosperms are vascular plants and are dominant sporophyte. Unlike ferns, gymnosperms produce seeds. The seeds of the gymnosperm are naked. This means that they are not encased or protected by an ovary. Most gymnosperms retain their green color all year long. They are **coniferous**, producing both male and female cones for reproduction. Common examples of gymnosperms are spruce, firs, and pines, but *Ginkgo biloba* is also included in this group. We will focus most of our attention on the pine. Pine trees have long, needle-like leaves, often in clusters. Most pine trees have both male pine cones and female pine cones. The male cones produce pollen and are smaller than female cones. Male cones are often grouped together at the ends of the pine's branches. They do not remain on the tree for long since they fall apart after their pollen is released.

We will look at the pine pollen under the microscope. The **pollen grain** stains dark while the two **air sacs** will appear light and almost transparent. The air sacs act like balloons to help the pollen grain to float in the wind. Female pine cones, on the other hand, are much larger than male cones. They also harden as they mature, a process than can take 1-3 years. You might remember putting glue and glitter on these types of cones in the winter or even covering them with peanut butter and birdseed to make bird feeders. Inside the scales of the female pine cones are the **ovules** or seeds. If you look closely in the scale, you will see two indentations where the two seeds were located. They have left the cone using their attached "wing" (inner piece of the scale).

© arka38/Shutterstock.com

Figure 14.5 Male and female gymnosperm cones.

Observe the pine cones in this photo and on display. How can you tell which one is male and which one is female?

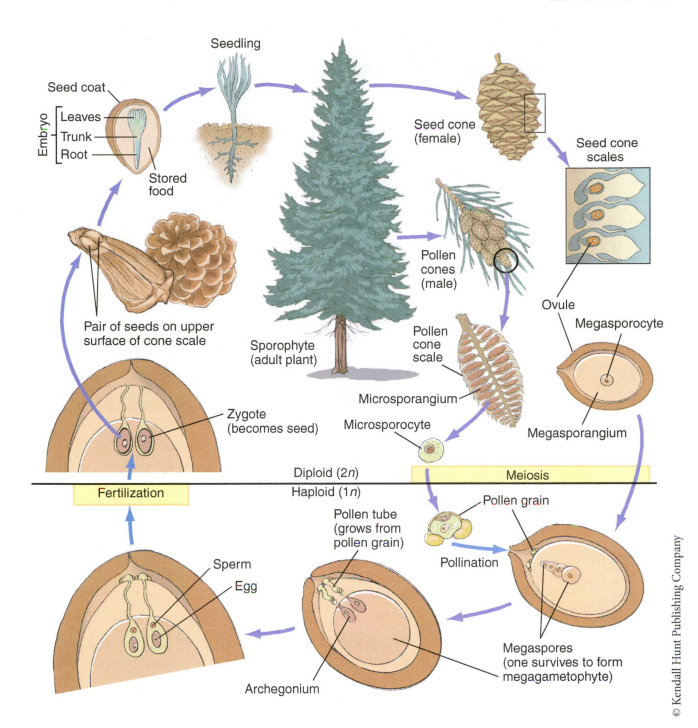

Figure 14.6 Conifer Life Cycle

Obtain a slide from your instructor. Focus the slide starting with the 4X objective and moving through each objective up to the 40X objective. Draw the specimen under the 40X objective.

Pine Pollen

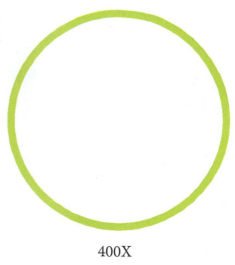

400X

QUESTIONS

6. What does the shape of the pine pollen resemble?

_____

7. What is the function of the air sacs surrounding a pollen grain?

_____

8. What is carried in pine pollen?

_____

ANGIOSPERM

Angiosperms are flowering plants. They have vascular tissue, are dominant sporophyte, and produce seeds that are protected in ovaries and produce flowers, and, fruit. There are many examples of angiosperms, such as daffodils, roses, apple trees, lilies, and grasses. We divide angiosperms into two major categories based on the structure of the **cotyledon** or seed leaf. The two categories are **monocot** and **dicot**. If the cotyledon is one single unit, the angiosperm is considered a monocot. When you look at a cross section of the stem of a monocot, the vascular tissue is scattered throughout, whereas in the root, the vascular tissue is in the form of a ring. Examples of monocots include grasses, grains, tulips, lilies, orchids, onions, and bamboo. In comparison, the cotyledon of a dicot is capable of being easily split into two equal parts. When cut in a cross section, a dicot's stem has vascular tissue arranged in a ring, whereas in the root, there is a clear arrangement of xylem in the shape of the letter "X." Examples of dicots are roses, buttercups, magnolias, poppies, and sunflowers.

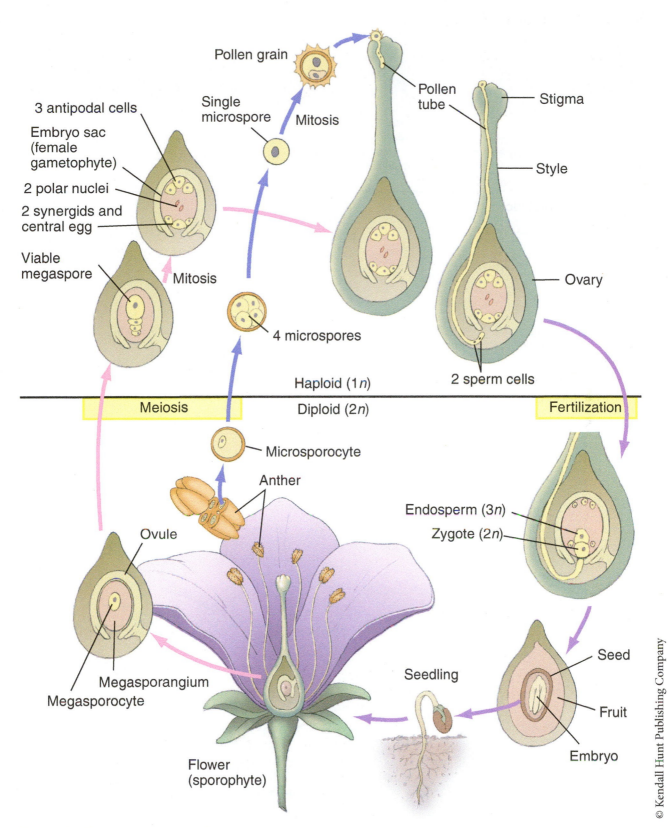

3 antipodal cells

Embryo sac
(female
gametophyte)

2 polar nuclei

2 synergids and
central egg

Viable
megaspore

Mitosis

Pollen grain

Single
microspore

Mitosis

4 microspores

Pollen
tube

Stigma

Style

Ovary

2 sperm cells

Haploid (1n)

Meiosis

Diploid (2n)

Fertilization

Microsporocyte

Anther

Ovule

Megasporangium

Megasporocyte

Flower
(sporophyte)

Endosperm (3n)

Zygote (2n)

Seedling

Seed

Fruit

Embryo

© Kendall Hunt Publishing Company

Figure 14.7 Angiosperm Life Cycle

Dicot (two cotyledons)

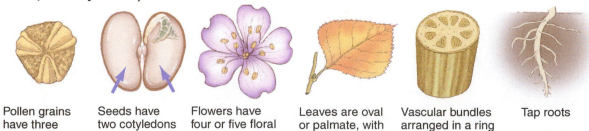

| Pollen grains have three pores or furrows | Seeds have two cotyledons | Flowers have four or five floral parts (or multiples thereof) | Leaves are oval or palmate, with net-like veins | Vascular bundles arranged in a ring around stem | Tap roots |

Monocot (one cotyledon)

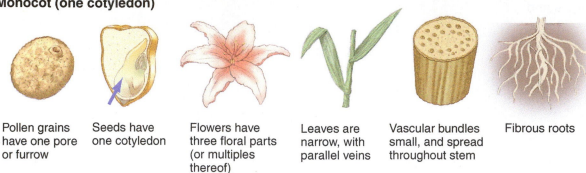

| Pollen grains have one pore or furrow | Seeds have one cotyledon | Flowers have three floral parts (or multiples thereof) | Leaves are narrow, with parallel veins | Vascular bundles small, and spread throughout stem | Fibrous roots |

Figure 14.8 Monocot and Dicot comparison

Flower Structure

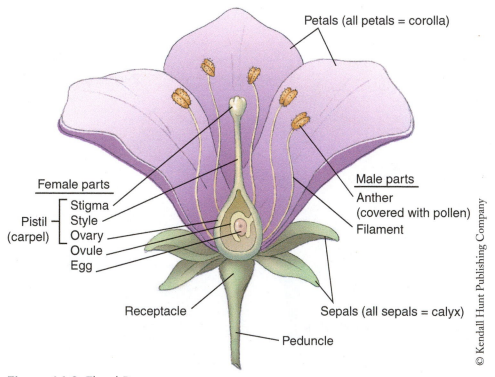

Petals (all petals = corolla)

Female parts
Pistil (carpel)
Stigma
Style
Ovary
Ovule
Egg

Male parts
Anther (covered with pollen)
Filament

Receptacle

Sepals (all sepals = calyx)

Peduncle

Figure 14.9 Floral Parts

Label the following structures on the flower model photo to the right (Figure 14.10).

- **Carpel** – female reproductive part of the flower
  - **Stigma** – the top of the carpel that sperm stick to
  - **Style** – sperm pierce through this long segment to get to the ovary
  - **Ovary** – contains the **ovules**; when matures becomes the fruit
- **Stamen** – male reproductive part of the flower
  - **Anther** – contains the pollen
  - **Filament** – stalk for the anther
- **Petals** – colored parts of the flower to attract pollinators
- **Sepals** – often green; located at the base of the flower below the petals

Figure 14.10 Flower Structure

QUESTIONS

7. Which part of the stamen contains pollen?

_____

8. Which part of the carpel is sticky?

_____

9. What color are the sepals in most angiosperms?

_____

Leaf Structure

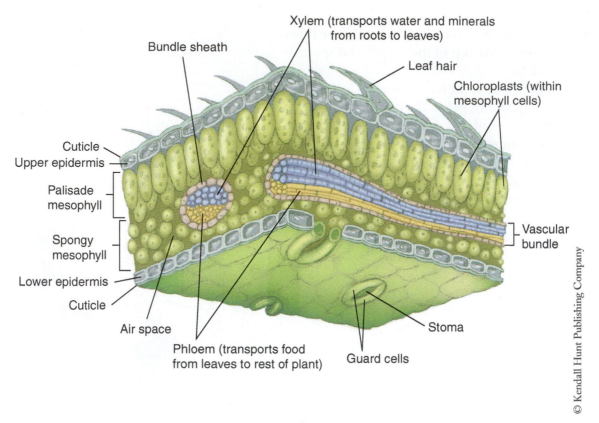

Figure 14.11 Dicot Leaf Cross-Section

© Kendall Hunt Publishing Company

Label the following structures on the leaf model photo on the next page (Figure 14.12).

- **Cuticle** – outermost covering of the leaf
- **Epidermis** – upper and lower cube-like cells between the cuticle and mesophyll
- **Mesophyll** – middle layers of cells
  - **Palisade mesophyll** – column shaped cells that have many chloroplasts
  - **Spongy mesophyll** – larger cells widely spaced
- **Vascular bundle** – veins covered with bundle sheath
  - **Xylem** – transports water
  - **Phloem** – transport sugar such as sucrose or sap
- **Stomata** – mouth-like structures that regulate the passage of gases from the air into the leaf
  - **Guard cells** – paired cells that open or close the stoma
  - **Stoma** – opening though which gases enter/leave the leaf

© Kendall Hunt Publishing Company

Figure 14. 12

QUESTIONS

10. What is the function of the guard cells?

_____

11. What substance is transported by the xylem?

_____

12. Name the dark green ovals inside each cell:

_____

13. What would accumulate in the spaces of the spongy mesophyll?

_____

Obtain a slide from your instructor. Focus the slide starting with the 4X objective and moving through each objective up to the 40X objective. Draw your specimen under the objective specified below.

Angiosperm Leaf

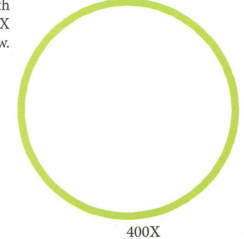

400X

14. Locate the palisade mesophyll on the slide. Describe the arrangement of the cells in this layer:

_____

Obtain a slide from your instructor. Focus the slide with the 4X objective and draw your specimen below. The dissecting scope can also be used.

Monocot Stem

40X

15. Describe the arrangement of vascular tissue in the monocot stem:

_____

Obtain a slide from your instructor. Focus the slide with the 4X objective and draw the specimen below. The dissecting scope can also be used.

Dicot Stem

40X

16. Describe the arrangement of vascular tissue in the diocot stem:

_____

Label each specimen a Monocot or Dicot

Stem Specimen A

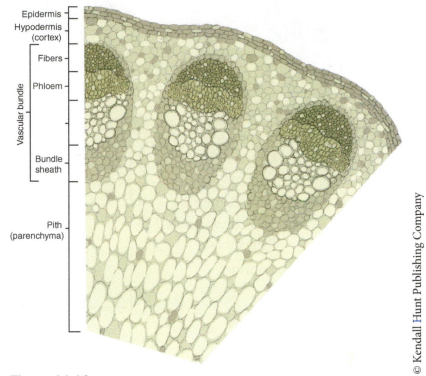

Figure 14.13

_____

Stem Specimen B

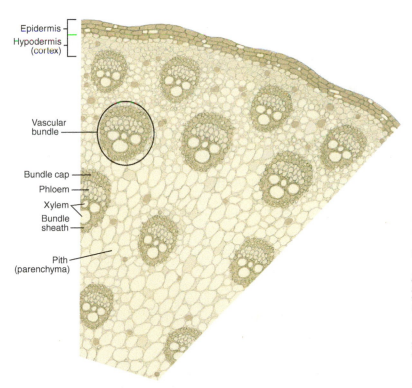

Figure 14.14

_____

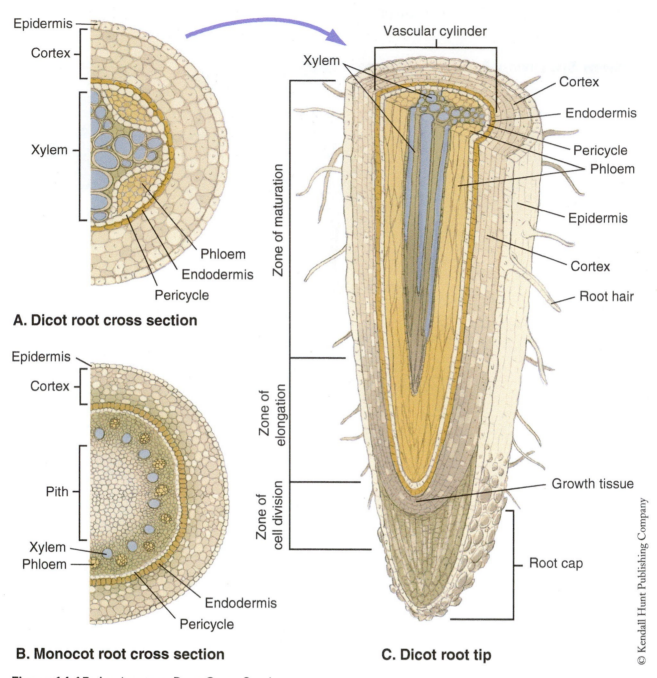

A. Dicot root cross section

B. Monocot root cross section

C. Dicot root tip

Figure 14.15 Angiosperm Root Cross-Section

© Kendall Hunt Publishing Company

Obtain a slide from your instructor. Focus the slide starting with the 4X objective and moving through each objective up to the 40X objective. Draw your specimen under the most clear, best viewed objective below.

Monocot Root

| 40X | 100X | 400X |

QUESTION

18. Describe the arrangement of vascular tissue in the monocot root:

_____

Obtain a slide from your instructor. Focus the slide starting with the 4X objective and moving through each objective up to the 40X objective. Draw your specimen under the most clear, best viewed objective below.

Dicot Root

| 40X | 100X | 400X |

QUESTION

19. Describe the arrangement of the xylem in the dicot root:

_____

LAB 15

Animals

The Animal Kingdom is composed of multicellular, eukaryotic, heterotrophic organisms. Animals must consume organic compounds made by other organisms. Most animal phyla are **invertebrates,** lacking a backbone, while others are **vertebrates** that have a backbone.

Recall the organization: Domain, Kingdom, Phylum, Class, Order, Family, Genus, and Species. When we refer to an organism, we will use its scientific name, a binomial, of genus and species. The genus is written first and capitalized. The species is written next and is in all lowercase. The entire binomial is written in italics such as *Homo sapiens* for a human. The following example is the classification for *H. sapiens*:

Domain: Eukarya
Kingdom: Animalia
Phylum: Chordata
Class: Mammalia
Order: Primates
Family: Hominidae
Genus: *Homo*
Species: *sapiens*

© L. Skywalker/Shutterstock.com

In classifying an animal, we look at the following criteria: symmetry, cephalization, segmentation, digestive tract, and germ layers.

Symmetry is the ability to divide an organism into parts using a median plane. There are three types of symmetry:

Asymmetrical – organism has no symmetry when cut in half, each half is unique in shape
Bilateral – adult organism can be divided equally into a right and left half where each half is a mirror image of the other
Radial – organism is organized around a central axis such as the spokes on a bike wheel

Cephalization is a concentration of nervous tissue at one end of the organism's body. In humans and other animals, this would be the brain.

Segmentation is often a difficult term to understand. It is the repetition of body parts along a longitudinal axis. In humans, this can be seen in the vertebral column.

Adapted from *BIO 105: Introduction to Biology Lab Manual* by Stephanie M. Brown. Copyright (c) 2009 by Kendall Hunt Publishing Company. Used with permission.

Digestive tracts can be either complete or incomplete. An organism with a **complete** digestive tract has two openings, one that takes in food and another opening that expels waste. An organism with an **incomplete** digestive tract has only one opening that both takes in food and expels waste.

Germ layers develop into a specific set of tissues (and organs, if present). There are three germ layers: ectoderm, mesoderm, and endoderm. Most animals have all three. To name a few examples: the **ectoderm** develops into the skin and nervous system, the **mesoderm** develops into muscle and bone, and the **endoderm** develops into the digestive organs.

H. sapiens have bilateral symmetry, cephalization, true segmentation, a complete digestive system, and three germ layers.

The following charts summarize the differences among the major groups we will study.

Observe the following slides and make detailed drawings. These slides may be set up as a demonstration.

Platyhelminth: Tapeworm (*Taenia pisiformis*) – observe hooks, suckers, proglottids

Table 15.1 Animalia Kingdom Summary

Phylum	Examples	Symmetry	Cephalization	Digestive Tract	Segmentation	Germ Layers	Coelomate	Notochord & Nerve Cord
Porifera	sea sponges, bath sponge, Grantia	asymmetrical	none	incomplete	none	none	Acoelomate	none
Cnidaria	jellies, coral, anemone, hydra	radial	none	incomplete	none	2	Acoelomate	none
Platyhelminths	planaria (*Dugesia*), tapeworm (*Taenia*), flukes	bilateral	yes	incomplete	none	3	Acoelomate	none
Nematoda	*Ascaris*, pinworms, hookworms, *Trichinella*, heartworms	bilateral	yes	complete	none	3	Pseudocoelomate	none
Annelida	earthworm, leech	bilateral	yes	complete	yes	3	Coelomate	none
Mollusca	clams, octopus, squid, chiton, snail	bilateral	yes (except bivalves)	complete	none	3	Coelomate	none
Arthropoda	bee, grasshopper, spider, horshoe crab, crayfish	bilateral	yes	complete	yes	3	Coelomate	none
Echinodermata	sea star, sand dollar, sea urchin	bilateral, radial as adults	none	complete	none	3	Coelomate	none
Chordata	tunicate, lancelet, fish, frog, snake, human	bilateral	yes	complete	yes	3	Coelomate	yes

Table 15.2 Chordata Phyla Summary

Phylum	Subphyla	Class	Examples	Symmetry	Cephalization	Digestive System	Segmentation	Germ Layers	Coelomate	Notochord & Nerve Cord
Chordata	Urochordata/Tunicates		sea squirts	bilateral	yes	complete	yes	3	Coelomate	yes
	Cephalochordata		lancelet *Amphioxus*	bilateral	yes	complete	yes	3	Coelomate	yes in larvae
	Vertebrata	Agnatha/jawless fish	lamprey, hagfish	bilateral	yes	complete	yes	3	Coelomate	yes
		Chondrichyes/cartilaginous fish	shark, sting ray	bilateral	yes	complete	yes	3	Coelomate	yes
		Osteichthyes/bony fish	perch, trout, goldfish	bilateral	yes	complete	yes	3	Coelomate	yes
		Amphibians	frog, salamander	bilateral	yes	complete	yes	3	Coelomate	yes
		Reptiles	birds, snake, crocodile, turtle, lizard, gecko	bilateral	yes	complete	yes	3	Coelomate	yes
		Mammals	humans, bat, cat, whale, kangaroo, anteater	bilateral	yes	complete	yes	3	Coelomate	yes

Nematoda: *Trichinella* – observe larvae coiled in a spiral embedded in muscle tissue

Total magnification: _____

Arthropoda: Spider or Insect – observe head and jointed appendages

Total magnification: _____

Observe the collection of Animalia in the lab and be able to classify the organism into its appropriate phyla, subphyla, and class where applicable. Many of these specimens are fragile so please take extra care. Answer the questions and key out the provided specimens to fill in the tables given.

QUESTIONS

1. Name one organism that has radial symmetry.

   _____

2. Name two organisms that are mollusks but do not have a shell outside their body.

   _____

3. Why does a tapeworm look as if it is segmented but it is not?

   _____

4. Define the term hermaphrodite. Why would this be advantageous for an animal?

   _____

   _____

   _____

   _____

QUESTIONS

5. Name two organisms that are hermaphrodites:

   _____

6. Why might a person have trouble moving and breathing if they have an infection with the roundworm *Trichinella*?

   _____

7. How is the spider different from other arthropods?

   _____

Use Key 1 to classify the provided vertebrate specimens to Class. List the common name of the organism, the steps necessary to key out the organism, and identifying characteristics of that class in Table 1.

KEY 1. SUBPHYLUM VERTEBRATA

1a.	Hair present	Class Mammalia
1b.	Hair absent	go to 2
2a.	Feathers present	Class Aves
2b.	Feathers absent	go to 3
3a.	Jaws present	go to 4
3b.	Jaws absent	Agnatha
4a.	Paired fins present	go to 5
4b.	Paired fins absent	go to 6
5a.	Skeleton bony	Class Osteichthyes
5b.	Skeleton cartilaginous	Class Chondrichthyes
6a.	Skin scales present	Class Reptilia
6b.	Skin scales absent	Class Amphibia

Name of Animal	Steps in Classification	Phylum	Subphylum	Class	Characteristics
TABLE 15.3 Vertebrates					
1.		Chordata	Vertebrata		
2.		Chordata	Vertebrata		
3.		Chordata	Vertebrata		
4.		Chordata	Vertebrata		
5.		Chordata	Vertebrata		
6.		Chordata	Vertebrata		
7.		Chordata	Vertebrata		
8.		Chordata	Vertebrata		
9.		Chordata	Vertebrata		
10.		Chordata	Vertebrata		
11.		Chordata	Vertebrata		
12.		Chordata	Vertebrata		

Use Key 2 to classify the provided invertebrate specimens to Phylum. If the organism is an arthropod, use Key 3 to derive class. If the organism is a mollusk, use Key 4 to identify class. List the common name of the organism, the steps necessary to key out the organism, and identifying characteristics of that phylum/class.

KEY 2. SELECTED INVERTEBRATE PHYLA

1a.   Body symmetry radial .... go to 2
1b.   Body symmetry not radial .... go to 3
2a.   Tentacles present, body soft ....Phylum Cnidaria
2b.   Tentacles absent, body hard and rough .... Phylum Echinodermata
3a.   Appendages (i.e. legs or tentacles) or a shell present... go to 4
3b.   Appendages (i.e. legs or tentacles) and a shell absent .... go to 5
4a.   Jointed legs present ....Phylum Arthropoda; **use KEY 3 ALSO**
4b.   Jointed legs absent...Phylum Mollusca; **use KEY 4 ALSO**
5a.   Body segmented, body round.. Phylum Annelida
5b.   Body not segmented, body flat .... Phylum Platyhelminthes

TABLE 15.4 Invertebrates				
Name of Animal	Steps in Classification	Phylum	Class	Characteristics
1.				
2.				
3.				
4.				
5.				
6.				
7.				
8.				
9.				
10.				
11.				
12.				

KEY 3. SELECTED CLASSES OF PHYLUM ARTHROPODA

1a.   Walking legs, more than five pairs .... go to 2
1b.   Walking legs, five or fewer pairs .... go to 3
2a.   Legs, one pair for each body segment .... Class Chilopoda
2b.   Legs, two pairs for each body segment .... Class Diplopoda
3a.   Antennae present .... go to 4
3b.   Antennae absent... Class Arachnida
4a.   Antennae, one pair.. Class Insecta
4b.   Antennae, more than one pair... Class Crustacea

KEY 4. SELECTED CLASSES OF PHYLUM MOLLUSCA

1a.   External shell visible .... go to 2
1b.   External shell not visible .... Cephalopoda
2a.   Shell, hinged........................... Bivalvia
2b.   Shell, not hinged .... Gastropoda

LAB 16

Population Ecology

The common idea of population would include examples like the student population at ECTC, the population of Kentucky, or the world's population of people. What does the word *population* mean? Actually in everyday conversation, when we use this word to refer to humans, "population" can refer to any group alive at a particular time in a particular place.

In ecology, however, the word *population* has a more restricted meaning. **A population is defined as members of the same species living at a particular time in a particular place and (theoretically) reproducing together.** Another way to put this is that members of a population share a group of genes, also known as a **gene pool**.

It can be difficult to determine where one population ends and another begins, and for this reason, most population research is carried out on islands or in other habitats with clear boundaries.

For example, one might study the population of bullfrogs in Freeman Lake.
What are two other examples of local ecological populations?

1. _____

2. _____

Since a population is always a group of individuals, it has group characteristics that are not found in single individuals. For example, a population has:

- a birth rate: the number of offspring born in a given time span
- a growth rate: the speed at which the population increases
- an age structure: the percentage of the population at each different age

What are two other characteristics that apply to a population, but not to an individual?

1. _____

2. _____

VARIATION AND AVERAGES IN A POPULATION EXERCISE

1. Does this class represent a population? _____
2. List five things that you can see are different in the students of this lab.
   a.
   b.
   c.
   d.
   e.

Adapted from *Laboratory Manual for the Processes of Life: BIO 101* by Holoyoke Community College. Copyright © 2013 by Kendall Hunt Publishing Company. Reprinted by permission.

3. Now choose something that can be easily measured to compare individuals in the class.
4. What is the average for the class (remember to use appropriate units)?
5. What is the variation for the class?
6. What factors could influence the variation?

7. On average, a person's wingspan (length from fingertip to fingertip when arms are outstretched) is expected to be the same as his or her height.

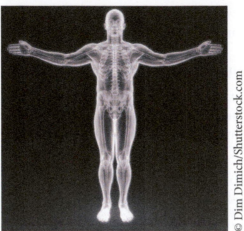

© Dim Dimich/Shutterstock.com

8. Do you think that your wingspan and height will be equal? _____
9. Calculate your ratio of wingspan to height (Wingspan/height).

   a. Wingspan _____

   b. Height _____

   c. Ratio _____
      A ratio of 1 is exactly the same, over 1 is longer wingspan and less than 1 is a shorter wingspan than height
10. Were you correct in your guess? _____
11. How much did you differ from an exact match? _____
12. Collect the wingspan, height, and ratio data from each student.

Name	Height	Wingspan	Ratio

Name	Height	Wingspan	Ratio

13. Now create a graph using each students wingspan versus height (remember to label axes).

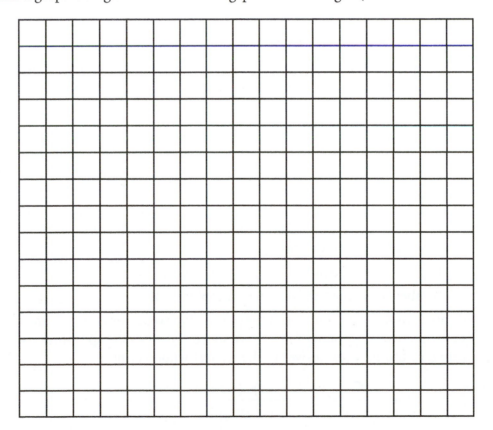

14. Calculate the average ratio of the class (add all ratios and divide by the number of students):
_____

15. Is this closer to 1 than your individual ratio?

16. What does this tell you about populations?

ESTIMATING POPULATION SIZE

One of the fundamental characteristics of any population is its size, how many members it contains. An exact count, or census, of a population often is very difficult to do, especially for large or mobile populations. For mobile populations, size can be estimated using the **Lincoln Index** or **Mark-Recapture** method. This method involves capturing a sample of animals (C), counting and marking them, releasing them back into their original area and allowing them time to disperse into the population (P), then capturing a second sample of animals (C_2). The fraction of marked individuals (R) in the second capture (R/C_2) is approximately equal to the fraction of marked individuals in the total population (C/P). In order for this to hold true, a few assumptions must be true. The experience of being captured before does not affect an animal's chance of being captured again (e.g., animals like getting captured, usually if food is available, or they avoid being trapped again) and the population has remained the same during the two sampling times (no emigration, immigration, births, or deaths). Working in groups, we will simulate this method using beads to represent animals.

Estimating Population Size Using the Lincoln Index

1. Each group should obtain one of the *populations of red beads* (animals) from your instructor. Each group also needs a *small container* (to simulate a net or live trap) and *some yellow beads*.
2. Fill a small beaker with red beads. This simulates your initial capture of the animals. Count the captured beads and write the result here:

$$C = \underline{\hspace{2cm}}$$

3. Set the captured red beads to one side. Count out an equal number of yellow beads and mix them thoroughly into the population (original container of red beads). This simulates marking the animals and releasing them back into the wild.
4. Pour beads from the container into the small beaker until it is completely full. Do not influence the color of the beads, pour out the beads randomly. A few yellow beads probably be mixed in with the red ones in your sample. Carry out these counts:

total number of beads in second capture = C_2 = _____

number of yellow beads in second capture = R = _____

5. As explained above, $R/C_2 = C/P$, where P is the population size. This equation may be rewritten to determine the estimate of the population size. Write your data in the appropriate places and carry out the following, rounding your answer to the nearest whole number:

$$P = \frac{C \cdot C_2}{R} = \frac{(\quad\quad) \cdot (\quad\quad)}{(\quad\quad)} =$$

6. Separate the yellow beads from the red, and return the beads to their original containers.
7. Write your data in the table below, and pass the materials on to the next group member and continue until all group members have estimated the population.

Group ID =

Name	Estimated Population Size
Group Average (GA) =	

8. Class Results and Conclusions. Put your group results on the board and copy the class data below.

Group ID	# Group Members	GA Estimated Population Size
A		
B		
C		
D		
E		
F		
Class Average =		

9. Your instructor will tell you the actual population size: _____

10. Looking at your group's data from #7, how consistent were the results? Was there much variability? What factors contributed to these results?

11. Looking at the class data from #8, what general conclusions can you draw about the accuracy of this method?

12. Is the class average a better estimate of the population size than some of the single trials? Why is it useful to carry out several trials?

13. Imagine that you were going to estimate the population size of the gray squirrels of the ECTC campus population by carrying out a Mark-Recapture technique. Gray squirrels mate in January–March and in June–July. Young are born 44 days after mating, on average. Given this information, what would be a good time of year to carry out your study? What would be a bad time of year? Why?

LAB 17

Simulated Disease Transmission

BACKGROUND

Infectious diseases are those which can be transmitted from a sick individual to an otherwise healthy individual. These diseases are caused by an infection from a pathogen that is now multiplying within the host's body while causing damage to it. Pathogens are any organisms or substances which cause disease by increasing in number within another organism and disrupting the normal functioning of that organism because of its growth. There are many bacteria, viruses, fungi, and eukaryotic parasites which fall into this category. These organisms are often referred to as microparasites because they are not visible to the naked eye. There can also be larger macroparasites which cause disease but are large enough to see with the naked eye; these include tapeworms, roundworms, and a variety of parasitic insect larva. There are many microorganisms growing on and within our body that do not interfere with our normal functioning and, in fact, can be beneficial. These beneficial symbiotic organisms are not considered pathogens.

Infectious diseases are very different than genetic diseases in that genetic diseases may be passed from parents to children, but never passed to someone other than to a direct offspring or descendent. Genetic diseases are not caused by a pathogen, but instead, specific disease alleles in the DNA of an organism.

Our body has a multi-leveled defense system known as our Immune System, but our defenses often need to encounter a specific pathogen before it recognizes it as being pathogenic. That is why we often

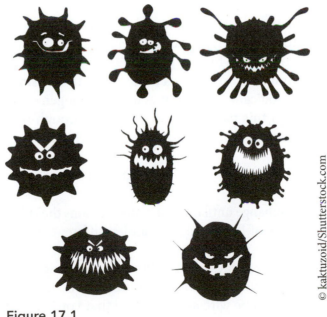

© kaktuzoid/Shutterstock.com

Figure 17.1

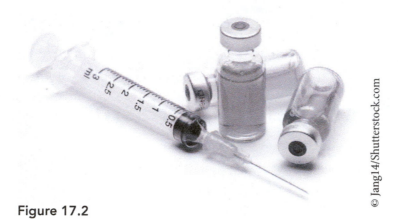

© Jang14/Shutterstock.com

Figure 17.2

get sick the first time we encounter a new pathogen, but can be resistant to it after that first illness. This is also the strategy behind vaccines. A vaccine is a collection of disease signature molecule (antigens), without having an active dangerous form of the pathogen present. The body is introduced to the signatures and marks it as dangerous so when a person is actually exposed to that pathogen, the immune system is already forewarned.

Infectious diseases are spread by the transfer of pathogens between individuals. This can occur through direct contact, fluid exchange (sexual and otherwise), indirect contact (touching the same surface that someone else had touched), aerosol droplets released by a cough or sneeze, contaminated food or drink, and insect vectors, just to name a few methods. Studying diseases and how they travel through a population is a specific field of biology known as epidemiology. In today's class we will not only see the spread of a simulated disease through our class population, but will also have the ability to track the path of infection for that disease.

Activity: Transmission of a Simulated Disease

Shared components

Simulated Disease Indicator solution.

1. In this lab activity, all participants will be given a tube of simulated body fluids. Upon receiving the tube, each student needs to use the pipette to transfer 1 mL of solution to a large test tube. Then discard the pipette, it will not be used further in today's activity. (Note: Before the simulation starts, there will be one vial of simulated body fluids that start with our simulated disease. The disease and body fluids are not biological and are not actually pathogens, so there is no need to worry about actually getting sick in this particular lab activity. Through the course of this lab activity other individuals will acquire the simulated disease. The original source and resulting infections will be chosen at random.)
2. It is important in labeling a sample of your simulated fluid before any exchanges take place.
3. **Rules for Fluid Exchange**
   a. Do not move on to fluid exchange until the instructor explains the rule for each round of exchange.
   b. There will be three rounds of fluid exchange and a different rule will precede each round of exchange.
   c. The rounds of exchange must always be simultaneous, meaning that everyone must find a partner and perform exchange round one before the entire group moves on to exchange round two.
   d. Each round of exchange must be with an individual that you did not exchange with in a previous round.

     e. Do not start the exchange of fluids until it is confirmed that everyone has found a legal partner for that round of exchange.

4. Fluid exchange in this activity consists of transferring ~3 mL of solution from the large tube of one participant into the large snap-cap tube of another participant. Mix thoroughly and then transferring ~3 mL of solution back to the first, followed by another round of mixing. Using a clean pipette, remove 1 mL of your fluid and place it in a test tube labeled with your initials and the exchange round (i.e., J.S. Round 1).

5. After completing the fluid exchange and recording the label of your exchange on the Student worksheet, return to your seat to await instructions on the next round of exchange.

6. After completing three rounds of fluid exchange, every participant will be tested for this simulated disease. The class will be invited to the front, one table at a time, for testing. Bring the large tube and your student worksheet.

7. Testing involves adding phenolphthalein to the large tube and mixing. If the tube remains clear, it is a negative response for this simulated disease, and you should return to your seat. If a pink color appears and the tube remains clear, that is a positive response to this simulated disease. You then need to record your label and the labels of your three exchanges on the class data table on the board.

8. All participants will be tested and the results on the board should be recorded on your student worksheet.

9. We will then attempt to determine the order of disease transmission by ruling out individuals who could not be the source. We do this by looking at the exchanges. If someone had an exchange with a person who ended up testing negative at the end of the experiment, we know that at the time of the exchange, that person did not have the disease and so they could not be the source, they did not get the disease until later exchanges.

10. Once we have narrowed down the possibilities of who the source could be, those individuals must bring their small tube up with them for a final round of testing.

Simulated Disease Transmission—Student Worksheet

Questions

1. Fill out the chart for your exchanges.

Exchange Round 1	Exchange Round 2	Exchange Round 3

2. Did you end up testing positive for the simulated disease?

3. Fill out the chart for the class data.

Number of Positive Tube	Exchange Round 1	Exchange Round 2	Exchange Round 3

4. Who was the source of our simulated disease?

Source	Exchange Round 1	Exchange Round 2	Exchange Round 3

5. Draw a flowchart of disease transmission, starting with our source and drawing a line to the names and showing who got infected for the first time in each round of exchange and from whom.

6. What causes infectious diseases?

7. Were you surprised by the number of people who tested positive by the end of the experiment? Why or Why not?

8. Why are individuals who tested negative on the data table?

9. How can one person spread a disease to so many other people?

10. What percentage of the participants in our class tested positive for the disease?

11. What would have happened if we would have had a fourth round of exchange?

LAB 18

Tree Identification Project

A tree is a perennial plant that typically has one trunk or stem that is at least 13 feet in height and at least 3 inches in diameter at breast height or approximately 4½ feet above the soil level. In North America, there are over 700 **native** tree species. The southeastern region of the United States has over 300 native tree species. Hundreds of different species from outside the continent are planted within the boundaries and referred to as **exotic** species. Trees are divided into two large categories for the purposes of identification: the conifers and the flowering trees. Conifers often possess cones rather than fruits, and the conifer leaf is usually evergreen. It lacks a stalk or petiole and has a shape like a needle or is scale-like. The leaves of flowering trees are usually deciduous, petioled, and have flattened blades. There some exceptions to these statements.

For example, the bald cypress is a deciduous conifer whose leaves resemble feathers; the Southern magnolia is a flowering tree with evergreen leaves and a cone-shaped fruit. When someone makes a collection of leaves, twigs, or fruits/cones of local common trees, that person becomes more observant of his or her natural surroundings.

Figure 18.1 Bald Cypress
Image courtesy of Betty Rosenblatt.

Figure 18.2 Southern Magnolia
Image courtesy of Betty Rosenblatt.

TREE CHARACTERISTICS AND HOW TO IDENTIFY A TREE SPECIES

Before you begin your collection, you will need to become familiar with common vegetative characteristics of trees. In flowering trees, the leaf may be simple with one blade or compound and the blade subdivided into leaflets. It will be important to notice the shape, the margin, the apex, and the base of the leaf or leaflet. Whether the tree has alternate, opposite, or whorled leaves is another characteristic to note. Look at the blades arrangement of veins.

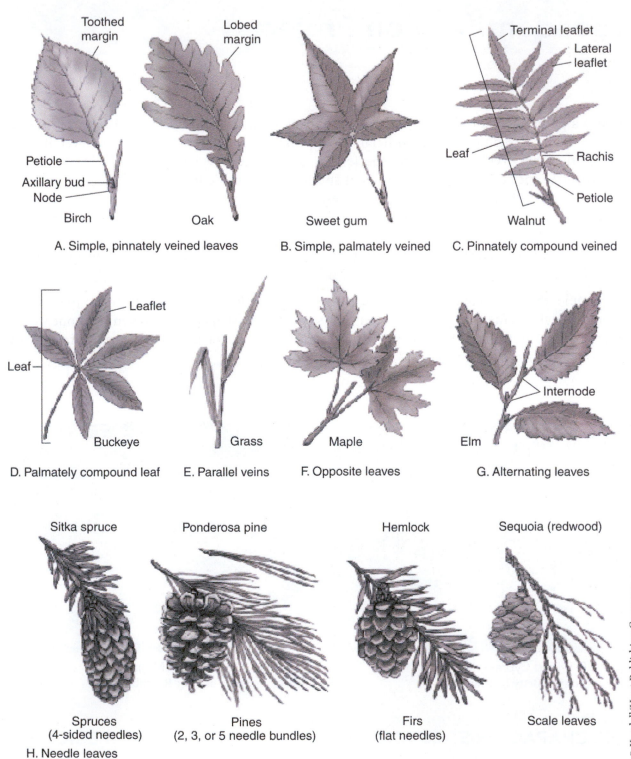

A. Simple, pinnately veined leaves   B. Simple, palmately veined   C. Pinnately compound veined

D. Palmately compound leaf   E. Parallel veins   F. Opposite leaves   G. Alternating leaves

Sitka spruce   Ponderosa pine   Hemlock   Sequoia (redwood)

Spruces (4-sided needles)   Pines (2, 3, or 5 needle bundles)   Firs (flat needles)   Scale leaves

H. Needle leaves

Figure 18.3 Leaf Types

© Kendall/Hunt Publishing Company

For the following, indicate the proper term descriptive of the tree's characteristics:

Simple or compound?

Alternate or opposite?

Palmately or pinnately netted?

Figure 18.4

Figure 18.5 Alternate or opposite?

Figure 18.6 Shape of leaves?

Simple or compound?

Figure 18.7

Images courtesy of Betty Rosenblatt.

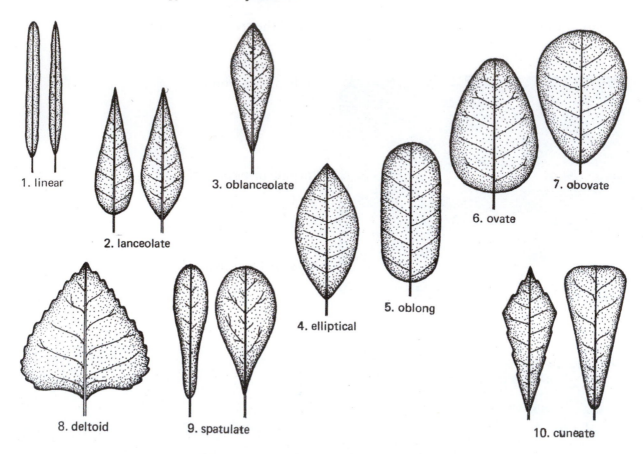

1. linear
2. lanceolate
3. oblanceolate
4. elliptical
5. oblong
6. ovate
7. obovate
8. deltoid
9. spatulate
10. cuneate

Figure A. Variation in General Outline

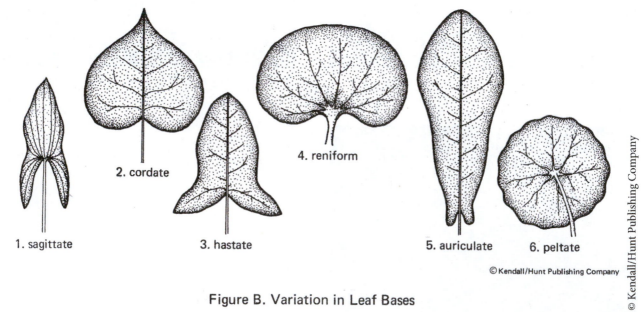

1. sagittate
2. cordate
3. hastate
4. reniform
5. auriculate
6. peltate

Figure B. Variation in Leaf Bases

Figure 18.8 Leaf Form Variations

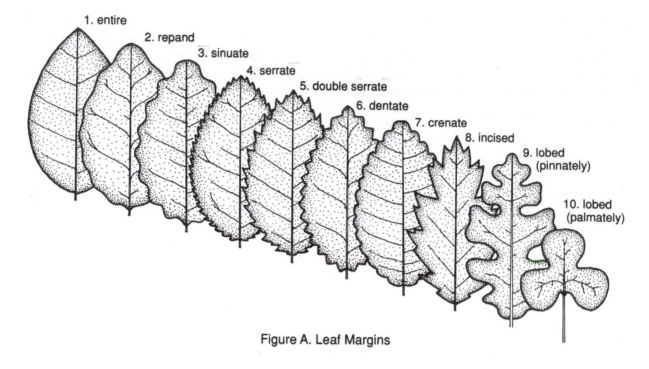

Figure A. Leaf Margins

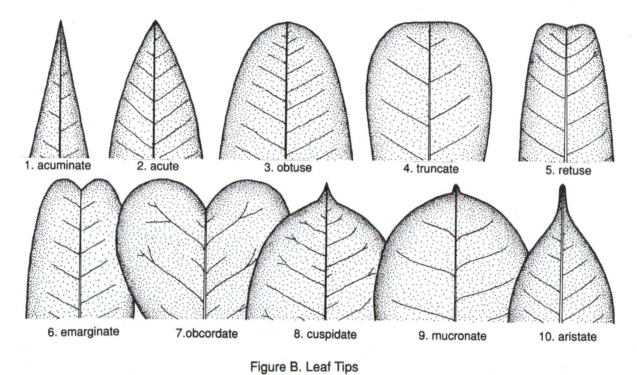

Figure B. Leaf Tips

Figure 18.9 Variations in Leaf Margins and Tips

It will be important to note the **shape**, the **margin**, the **apex**, and the **base** of the leaf or leaflet. Refer to Figure 2 "Leaf Form Variations" and Figure 3 "Variations in Leaf Margins and Tips" for help with these variations. After you feel comfortable with the terms descriptive of tree characteristics, you will want to locate **tree identification guides** in the library or on the Internet. Your instructor may give you recommended resources to help with the identification of local species.

A **dichotomous key** is very helpful. To use this kind of key, you will read pairs of descriptions (for example: 1. Leaf needle-like or scale-like or 1. Leaf not needle-like or scale-like, but a flattened single blade or leaflets) and select the one that fits your specimen. With each pair, you will make one selection and continue to follow the path until you receive the identity of your specimen tree. You will find many of these interactive keys on the Internet. To help you get started with tree identifications, go to this Internet site, http://www.cas.vanderbilt.edu/bioimages/tree-key/tree-key.htm, and key out the tree species below. Give both the common name and the scientific name.

Figure 18.10

Figure 18.12

Figure 18.11

Images courtesy of Betty Rosenblatt.

COLLECTING THE SPECIMENS

You should collect at least 25 leaf specimens from native trees. You may also collect native shrubs if the instructor allows you to include them, but you should focus mainly on collecting leaves from tree species. Remember that native trees and shrubs are those that grow naturally in the wooded habitats of your region. You may also find many of these trees and shrubs in your own neighborhood in yards, parks, and school grounds. Native trees and shrubs fare much better in any type of regional landscape than do cultivated plants brought in from a different region because their native gene pools are adapted for survival in the region. Do not collect cultivated plants acquired from plant nurseries. The list in Part IV will help you to determine whether you should include a specimen in the collection. There are some introduced species that are reproducing and that biologists consider naturalized species. You may include these in the collection. Ask your instructor if you are not certain, or simply collect extra specimens. Remember that you need at least 25 correctly identified native species for your scrapbook collection.

Before you make collections, check out a plant press from the instructor or make your own from cardboard, cut to size, and two pieces of plywood to sandwich the ten to twenty cardboards that will hold your leaf specimens for drying and pressing. To apply the necessary pressure to flatten your leaf specimens between the press, you will need a pair of straps or some mechanism to keep the pressure applied to the plant press. Once you have a press, you are ready to assemble the other materials you will need: newspapers, plant shears, notepad, tree key/guide.

Start collecting. Place the leaf specimens in the press as soon as possible between a folded newspaper sheet; then place them between a pair of the cardboards. You can place several specimens in separate newspaper sheets between a pair of the cardboards. As you collect each specimen, number the newspaper sheet and record data in your notepad for this specimen number. Record the following: name of specimen, collection site, date. If you cannot identify the specimen, try again later, but record as much information as you can. Use the Internet as an additional resource to aid in identification. Another good resource is http://www.eNature.com. You can type in your zip code and find color images of most of the local tree species. Woody Plants in North America, a multimedia CD set by John R. Seiler, John Peterson, and Edward C. Jensen published by Kendall Hunt, is the best resource of 860 trees, shrubs, and vines of North America.

IMPORTANT NOTE: You must cut the specimen's stem below the points of attachment (node) of at least two leaves. As you place the specimen between the newspaper sheet, turn one of the leaves or a leaflet, if compound, to show the lower surface. It is often necessary to observe the lower leaf surface as well as the upper leaf surface for correct identification as to exact species name. Remember to keep the specimen size appropriate to fit the scrapbook page you plan to display the specimen on once it is completely dried. Keep in mind that your goal is not only to press the leaves, but to get them dry as quickly as possible. Don't place wet leaves in the press. They will become moldy and unacceptable for your scrapbook display. Place the press in a sunny window, if possible.

DISPLAYING THE PRESSED COLLECTION

After the leaf specimens are completely dry, remove them from the plant press. You are now ready to prepare each leaf specimen for presentation format in your collection notebook. The specimen should fit the page and not have any parts extending beyond the page. Remember that you must display both leaf surfaces.

To secure the leaf specimen to the page (this should be thick paper), brush a diluted white glue mixture (1 part glue, 1 part water, and a squirt of liquid soap) onto the back of the leaf specimen. Be careful not to tear the dried specimen. Keep the layer of glue mixture thin. Carefully place the specimen onto the page, and place wooden blocks (or something that will hold the specimen down) on the top of the specimen until dried.

Place the following information in the lower right hand corner of the specimen sheet.

Common Name
Scientific Name
Collection Site and Habitat Type
Date of Collection
Name of Collector

You might wish to laminate the page, or to place it into a report cover for added protection.

LIST OF REGIONAL TREES AND SHRUBS OF THE MID-SOUTH

Common Name	Scientific Name
Eastern Red Cedar	*Juniperus virginiana*
Bald Cypress	*Taxodium distichum*
Northern White Cedar	*Thuja occidentalis*
Shortleaf Pine	*Pinus echinata*
Eastern White Pine	*Pinus strobus*
Loblolly, Yellow Pine	*Pinus taeda*
Virginia Pine	*Pinus virginiana*
Eastern Hemlock	*Tsuga canadensis*
Yew	*Taxus canadensis*
Ginkgo	*Ginkgo biloba*
Box Elder Maple	*Acer negundo*
Red Maple	*Acer rubrum*
Silver Maple	*Acer saccharinum*
Sugar Maple	*Acer saccharum*
American Smoketree	*Cotinus obovatus*
Winged Sumac	*Rhus copallinum*
Smooth Sumac	*Rhus glabra*
Poison Ivy	*Toxicodendron radicans*
Pawpaw	*Asimina triloba*
American Holly	*Ilex decidua*
Winterberry	*Ilex verticillata*
Devil's Walking Stick, Hercules' Club	*Aralia spinosa*
Smooth Alder	*Alnus serrulata*
River Birch	*Betula nigra*
American Hornbeam	*Carpinus caroliniana*
Eastern Hophornbeam	*Ostrya virginiana*
Catalapa	*Catalapa bignonioides*
Elderberry	*Sambucus canadensis*
Blackhaw	*Viburnum prunifolium*
Eastern Burningbush or Hearts'-a-bustin'	*Euonymus atropurpurea*
Strawberry Bush	*Euonymus americana*

Alternate-leaf Dogwood	*Cornus alternifolia*
Roughleaf Dogwood	*Cornus drummondii*
Flowering Dogwood	*Cornus florida*
Water Tupelo	*Nyssa aquatica*
Black Gum	*Nyssa sylvatica*
Persimmon	*Diospyros virginiana*
Sourwood	*Oxydendrum arboreum*
Wild Azalea	*Rhododendron canescens*
Sparkleberry	*Vaccinium arboreum*
Mimosa	*Albizia julibrissin*
Eastern Redbud, Judas Tree	*Cercis canadensis*
Honey Locust	*Gleditsia triacanthos*
Kentucky Coffeetree	*Gymnocladus dioicus*
Black Locust	*Robinia pseudoacacia*
Yellowwood	*Cladrastis kentuckea*
American Chestnut	*Castanea dentate*
American Beech	*Fagus grandifolia*
White Oak	*Quercus alba*
Scarlet Oak	*Quercus coccinea*
Southern Red Oak	*Quercus falcata*
Shingle Oak	*Quercus imbricaria*
Overcup Oak	*Quercus lyrata*
Bur Oak	*Quercus macrocarpa*
Blackjack Oak	*Quercus marilandica*
Basket or Swamp Chesnut Oak	*Quercus michauxii*
Chinkapin Oak	*Quercus muehlenbergii*
Water Oak	*Quercus nigra*
Pin Oak	*Quercus palustris*
Willow Oak	*Quercus phellos*
Northern Red Oak	*Quercus rubra*
Shumard Oak	*Quercus shumardii*
Post Oak	*Quercus stellata*
Nuttall Oak	*Quercus nuttallii*
Black Oak	*Quercus velutina*
Virginia Willow	*Itea virginica*
Sweetgum	*Liquidambar syraciflua*
Dwarf Red Buckeye	*Aesculus pavia*
Wild Hydrangea	*Hydrangea arborescens*
Mockernut Hickory	*Carya tomentosa*
Pignut Hickory	*Carya glabra*
Pecan	*Carya illinoinensis*
Shellbark Hickory	*Carya laciniosa*
Shagbark Hickory	*Carya ovata*
Butternut	*Juglans cinerea*
Black Walnut	*Juglans nigra*
Spicebush	*Linera benzoin*
Sassafras	*Sassafras albidum*

Tulip Tree, Tulip Poplar	*Liriodendron tulipifera*
Southern Magnolia	*Magnolia grandiflora*
Bigleaf Magnolia	*Magnolia macrophylla*
Umbrella Magnolia	*Magnolia tripetala*
Sweetbay Magnolia	*Magnolia virginiana*
Osage-orange	*Maclura pomifera*
White Mulberry	*Morus alba*
Red Mulberry	*Morus rubra*
White Ash	*Fraxinus americana*
Green Ash	*Fraxinus pennsylvanica*
American Sycamore	*Platanus occidentalis*
Carolina Buckthorn	*Rhamnus caroliniana*
Downy Serviceberry, Sarvisberry	*Amelanchier arborea*
Hawthorn	*Crataegus crus-galli*
Crabapple	*Malus angustifolia*
Apple	*Malus pumila*
Chickasaw Plum	*Prunus angustifolia*
Mexican Plum	*Prunus mexicana*
Peach	*Prunus persica*
Black Cherry	*Prunus serotina*
Pear	*Pyrus communis*
Buttonbush	*Cephalanthus occidentalis*
Eastern Cottonwood	*Populus deltoidus*
Black Willow	*Salix nigra*
Weeping Willow	*Salix babylonica*
Empress Tree, Paulownia	*Paulownia tomentosa*
Tree of Heaven	*Ailanthus altissima*
Snowbell	*Styrax americanus*
Basswood	*Tilia americana*
Hackberry, Sugarberry	*Celtis laevigata*
Water Elm	*Planera aquatica*
Winged Elm	*Ulmus alata*
American Elm	*Ulmus Americana*
Slippery Elm	*Ulmus rubra*
American Beautyberry	*Callicarpa americana*